Collector's Guide to the MICA GROUP

Schiffer Earth Science Monographs Volume 1

4880 Lower Valley Road, Atglen, Pennsylvania 19310

Robert J. Lauf

Schiffer Books are available at special discounts for bulk purchases for sales promotions or premiums. Special editions, including personalized covers, corporate imprints, and excerpts can be created in large quantities for special needs. For more information contact the publisher:

Published by Schiffer Publishing Ltd.
4880 Lower Valley Road
Atglen, PA 19310
Phone: (610) 593-1777; Fax: (610) 593-2002
E-mail: Info@schifferbooks.com

Please visit our web site catalog at **www.schifferbooks.com**

We are always looking for people to write books on new and related subjects. If you have an idea for a book, please contact us at the above address.

This book may be purchased from the publisher.
Include $5.00 for shipping.
Please try your bookstore first.
You may write for a free catalog.

In Europe, Schiffer books are distributed by:
Bushwood Books
6 Marksbury Ave.
Kew Gardens
Surrey TW9 4JF
England
Phone: 44 (0)208 392-8585
Fax: 44 (0)208 392-9876
E-mail: Info@bushwoodbooks.co.uk

Website: www.bushwoodbooks.co.uk
Free postage in the UK. Europe: air mail at cost.
Try your bookstore first.

Other Schiffer Books by Robert J. Lauf
Introduction to Radioactive Minerals.

Other Schiffer Books on Related Subjects
Collecting Fluorescent Minerals. Stuart Schneider
The World of Fluorescent Minerals. Stuart Schneider

Library of Congress Control Number: 2008924329

Designed by Mark David Bowyer
Type set in Arno Pro / Humanist 521 BT

ISBN: 978-0-7643-3047-6
Printed in China

Contents

Preface

The idea for this volume began with a paper published by the author in *Rocks & Minerals* (Lauf 2006) as part of a series of articles on important groups of so-called rock forming silicates, the purpose of which is to help mineral collectors gain a better appreciation of these complex minerals. Because of the importance of rock forming minerals in geological processes, they are the subject of extensive published research, much of which has been brought together in the five-volume compendium *Rock-Forming Minerals* (Deer, Howie, and Zussman 1962) and the greatly expanded Second Edition thereof. Within the mineral collecting community, however, rock-forming minerals as a group rarely get the attention they deserve. Why is that? First, their chemistry and, by extension, their nomenclature is more complicated than that of minerals with simpler composition such as calcite. Second, many cannot easily be distinguished from others in their group with the naked eye or by simple tests. Third, the opinion is sometimes expressed that "rock forming minerals rarely make nice crystals." This volume is an effort to address all three of these concerns.

This monograph is organized as follows: After a brief introduction, the general treatment begins with an explanation of the chemistry and taxonomy of the group with supporting tables to illustrate the logic of their classification and relationships to one another. A section on their formation and geochemistry explains the kinds of environments where micas can form. Then, a detailed entry for each mineral provides extensive locality information and full-color photos wherever possible so that collectors can see what good specimens look like. The final sections highlight additional topics of interest to collectors, including attractive associations of micas with other minerals, pseudomorphs involving mica, and micas that are fluorescent under UV light.

Acknowledgments

Vandall King provided a critical review of the manuscript and made numerous constructive comments and suggestions. The following colleagues kindly provided technical information, literature, and helpful discussions: Petr Černy, *University of Manitoba*; Deborah Cole, *Oak Ridge National Laboratory*; Frank Hawthorne, *University of Manitoba*; Lance Kearns, *James Madison University*; and Tom Watkins, *Oak Ridge National Laboratory*. Important specimens and background information were supplied by: Dave Bunk; Isaias Casanova, *IC Minerals*; Sharon Cisneros, *Mineralogical Research Co.*; Richard Dale, *Dale Minerals*; Shields Flynn, *Trafford-Flynn* Minerals; Beau Gordon, *Jendon Minerals*; Leonard Himes, *Minerals America*; Rob Kulakofski, *Color-Wright*; Tony Nikischer, *Excalibur Mineral Co.*; Neal Pfaff, *M. Phantom Minerals*; C. Carter Rich; Jeff Schlottman, *Crystal Perfection*; Jaye Smith, *The Rocksmiths*; Sergey Vasiliev, *Systematic Mineralogy*; and Chris Wright, *Wright's Rock Shop*.

Introduction

The micas are a group of about forty pseudohexagonal monoclinic silicate minerals characterized by a molecular sheet structure and a perfect basal cleavage. Many of the species are common rock-forming minerals and occur in igneous, metamorphic, and sedimentary deposits. The group offers many attractive examples worthy of a place in any general, systematic, or display collection. For the advanced collector, the mica group offers rich opportunities to assemble a specialty collection or display along a number of interesting themes.

Mica has been an important industrial commodity for many years. The earliest application, as furnace windows, made use of its properties of heat resistance and transparency, along with its ability to be cleaved into thin yet strong and flexible sheets. Thin sheets of mica were also used as insulators in vacuum tubes and capacitors. Presently the largest use of natural mica is in weather-resistant roofing materials and other construction products, including exterior paints. Finely ground micas are used in pearlescent pigments and cosmetics, and since the 1960s virtually all micas in cosmetic products are synthetic. Early attempts to make synthetic micas focused on duplicating the compositions of natural micas. However, most natural micas contain the hydroxyl ion [for example, the nominal composition of phlogopite is $KMg_3AlSi_3O_{10}(OH)_2$] and in order to create this phase synthetically, high pressure is needed to prevent decomposition of the structure through loss of water. This problem was eliminated by the development of synthetic fluoro-micas, in which F^- replaces the OH^- ion [the nominal end-member composition of fluorophlogopite would be $KMg_3AlSi_3O_{10}F_2$]. The fluoro-micas are stable enough to be prepared from the melt at ordinary pressures. Glass-ceramic products such as cookware and machinable insulating materials rely on microscopic crystallites of mica for their excellent properties. Mica is also used in various insulating materials, electronic circuit boards, and paper products. It is a measure of the industrial value of micas that many hundreds of patents have been granted for methods of making micas and products containing them. Interestingly, natural micas whose compositions lie within the field of fluorophlogopite were only found recently and fluorophlogopite was approved as a new species in 2006.

Members of the mica group are generally too soft to be cut as gems or fashioned into jewelry by themselves; however, several lapidary and ornamental stones get their color from inclusions of mica. The term *aventurine* is applied to quartz that contains sparkling inclusions of green chromian muscovite var. *fuchsite*, and the optical phenomenon is generally referred to as aventurescence when a transparent mineral has shiny flakes of any included mineral. Green muscovite disseminated in marble is called *mariposite* from the locale in Mariposa County, California. The term *mariposite* has in the past also been applied informally to the chromian muscovite itself, particularly for material from the California locality; used in that sense it would be synonymous with *fuchsite*. Mariposite rock is quarried as a decorative stone for landscaping, and is also fashioned into cabochons, spheres, and other decorative objects. Compact lavender/pink lepidolite (sometimes containing needles of deep pink tourmaline) is sometimes carved or slabbed to make paperweights, bookends, and the like. It is soft enough to be worked with steel tools but is difficult to polish well because of undercutting, particularly when inclusions of harder minerals are present. Jewelry containing cabochons of this material are occasionally seen in the lapidary trade.

Figure 1. A slab of aventurine (green chromian muscovite on quartz) from India. This material is commonly used to make pleasing green cabochons. RJL3179

Figure 2. A simple carving of a cat about 6 cm tall, made from green aventurine.

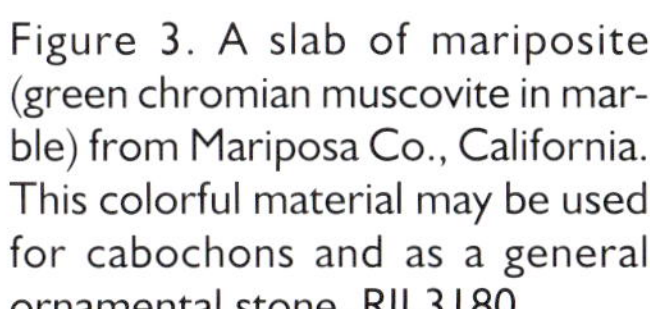

Figure 3. A slab of mariposite (green chromian muscovite in marble) from Mariposa Co., California. This colorful material may be used for cabochons and as a general ornamental stone. RJL3180

Figure 4. Slabs of massive purple lepidolite from Pala, California. This material is suitable for making cabochons, although it tends not to take as good a polish as many traditional semiprecious stones. RJL3181

Taxonomy of the Mica Group

The micas form a large group that covers a considerable range of compositions, many of which form continuous solid solution series with others. Not surprisingly, organizing the group into a rational set of accepted mineral species is a painstaking process. The classic work *Rock-Forming Minerals* (Deer, Howie, and Zussman 1962) described nine "common" species: muscovite, paragonite, glauconite, phlogopite, biotite, lepidolite, zinnwaldite, margarite, and clintonite, while recognizing that the nomenclature of the group as a whole was imperfect. A decade ago, the Commission on New Minerals and Mineral Names of the International Mineralogical Association appointed a subcommittee to establish recommended nomenclature for this important group based on modern understanding of chemistry and crystal structure relationships (Rieder et al. 1998). Their recommendations also appear in abbreviated form in a recent edition of *Fleischer's Glossary of Mineral Species* (Mandarino 1999). The second edition of *Rock-Forming Minerals* (Fleet 2003) closely follows current nomenclature and treats the individual mica species in significantly greater detail.

General formula

Micas are represented by the following simplified formula:

$\mathbf{I}\mathbf{M}_{2\text{-}3}\square_{1\text{-}0}\mathbf{T}_4\mathrm{O}_{10}\mathbf{A}_2$ where:

I (interlayer cations) is K, Na, Ca, or (less-commonly) Cs, NH_4, Rb, or Ba;

M (octahedral cations) is generally Li, Fe^{2+}, Fe^{3+}, Mg, Al, Ti or (less often) Mn^{2+}, Mn^{3+}, Zn, Cr, or V;

$\square$ represents a vacancy;

T (tetrahedral cations) is commonly Si, Al, Fe^{3+}, or (less often) B or Be;

A is commonly F, OH, or (less often) O, Cl, or S.

As with other complex silicates, the basis for assigning species to groups or subgroups and assigning a particular specimen to one of the recognized species involves understanding which ions occupy particular structural positions. Based on how many of the interlayer (**I**) ions are divalent vs. monovalent, the micas are subdivided into true micas, Table 1, (if >50% of the **I** cations are monovalent) and brittle micas, Table 2, (if >50% of the **I** cations are divalent). A third group, the "interlayer-cation-deficient micas", is of somewhat less interest to most collectors, but for completeness these species are listed in Table 3. These groups are further subdivided into dioctahedral and trioctahedral micas, based on occupancy of the octahedral (**M**) sites. The ions occupying these three sites contribute a net +6 charge per formula unit, but one can see from the formula that the **M** ions can be divalent (e.g, Fe^{2+}, Mg, Mn^{2+}), trivalent (e.g., Fe^{3+}, Cr, Al), or even monovalent (Li). Thus, if the **M** ions are mostly divalent there will be about three of them, whereas if the **M** ions are mostly trivalent there would only be about two and the third site would be vacant. So for classification purposes, the dioctahedral micas are defined by having less than 2.5 octahedral (**M**) cations and trioctahedral micas by having more than 2.5 octahedral (M) cations per unit cell.

Crystal structure and morphology

The crystal structure of mica consists of "2:1 layers" having a net negative charge, which are compensated and bonded together by large, positively charged "interlayer cations." The 2:1 layer consists of two tetrahedral sheets with an octahedral sheet sandwiched between them. Within the tetrahedral sheet, the individual tetrahedra share three corners each with adjacent tetrahedra, giving rise to a hexagonal mesh pattern. The fourth corner (referred to as the "apical oxygen") points in a direction perpendicular to the plane of the sheet and is, in turn, part of the adjacent octahedral sheet, made up by individual octahedra sharing octahedral edges.

Micas are divided into true micas and brittle micas based on the layer charge (x) per formula unit. In the true or "flexible" micas, the layer charge is ideally x = -1, so electrostatic compensation involves monovalent interlayer cations such as K^+ and Na^+. In the brittle micas Al^{3+} substitutes for Si^{4+} on some of the tetrahedral sites, ideally changing the layer charge to x = -2, so compensation is provided mainly by divalent interlayer cations, usually Ca^{2+} or Ba^{2+} (Bailey 1984). In the brittle micas, the higher charge on the 2:1 layers (-2 versus -1 in the true micas) creates a stronger bond between the interlayer cations and adjacent oxygens. This limits the flexibility of the mica layers and explains their generally brittle behavior. From the ideal end-member formulas one can see that the brittle mica margarite is related to muscovite by the substitution of Ca^{2+} for K^+ and similarly the brittle mica clintonite is the Ca^{2+} analogue of phlogopite.

Within each 2:1 layer the upper tetrahedral sheet is staggered relative to the lower tetrahedral sheet to provide coordination around the medium-sized octahedral cations. This stagger may be directed positively or negatively along any of the three crystallographic directions of the pseudohexagonal network. If the direction of stagger varies regularly from layer to layer, then the crystal structure will repeat itself along the *Z* direction in some multiple of 10Å and the mica will display *polytypism*. Layer structures that are substantially the same composition and differ only in the layer sequences are called *polytypes* and will have different crystallographic symmetries. The different possible arrangements give rise to six standard polytypes, five of which have been found in natural micas. Some tend to be more abundant than others and most mica species favor particular polytypes. In the individual mineral entries that follow, reported polytypes will be listed, with **boldface** indicating the most commonly observed polytype(s) for each species if more than one polytype is known. Although polytypism represents a fine detail of crystal structure that is of limited interest to the mineral collector, it is occasionally important to geochemists because energetic differences between some polytypes can provide clues to the temperature at which the mica crystallized.

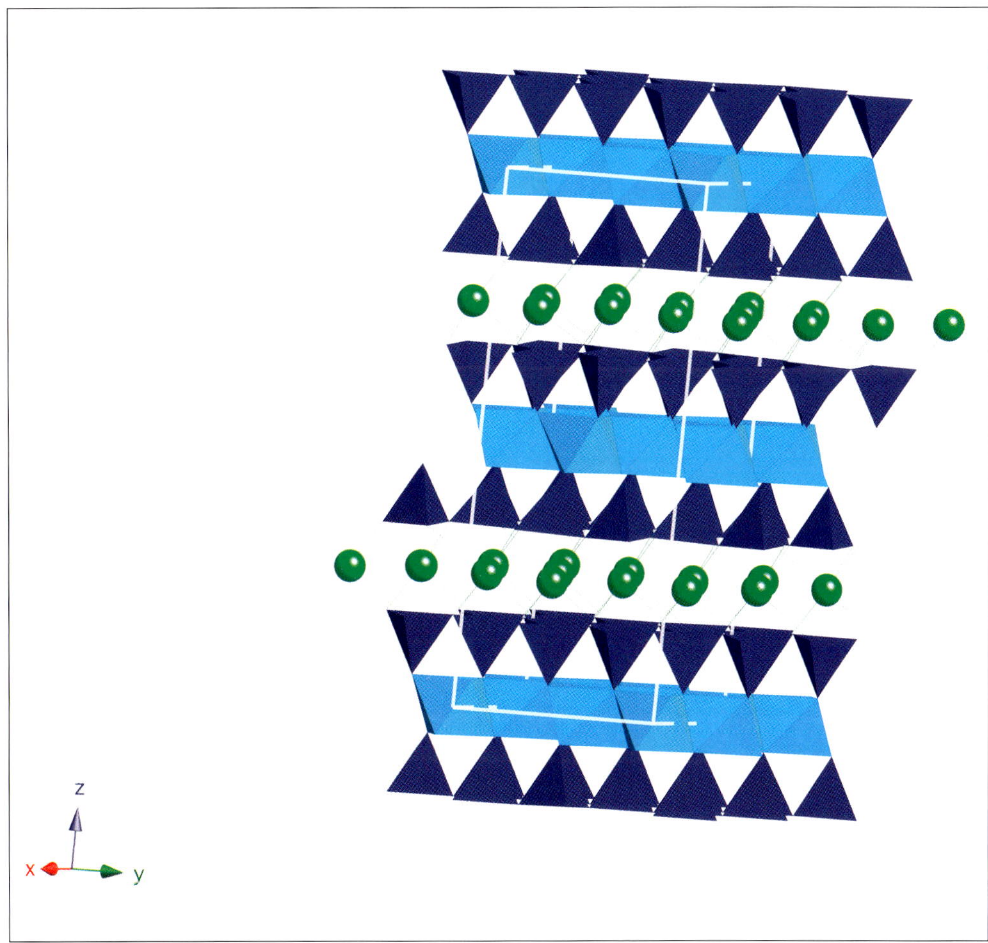

Figure 5. The crystal structure of muscovite viewed parallel to the cleavage plane. In this type of model, each coordination polyhedron represents a metal ion (hidden within the polyhedron) surrounded by oxygen ions at the corners. Dark blue tetrahedra represent (SiO_4) groups in layers on either side of an octahedral layer shown in light blue. The "2:1" layers are in turn bonded to the K^+ interlayer ions shown in green.

Although the word mica often brings to mind thin, flat sheets with perhaps simple hexagonal outline, the mica group presents a rich diversity of crystal forms and habits. Mica that has grown into a cavity can form large yet exquisitely thin tabular crystals as well as more equant shapes. Large crystals showing a number of interesting forms can also be found solidly encased in marble or carbonatite; with judicious cleaning or trimming these can make superb specimens. Goldschmidt (1918) illustrated morphologies for 131 examples of true micas ("Glimmer-Gruppe") and 24 examples of brittle micas ("Sprödglimmer").

When presented with a crystal that has a perfectly smooth, flat "termination" it is natural to wonder whether the crystal really formed such a smooth surface or was instead cleaved during collection or trimming. In many cases one can never be sure, but a few observations might be made. First, examine the whole specimen. If the opposite end of the crystal (or the ends of other crystals) is still partly encased by the matrix and it, too, is quite flat, then it is possible that the exposed end is a natural termination. Second, examine the partly exposed end to look for finer structural details such as luster or the presence of very small surface figures or growth steps. If the questionable face is significantly smoother and shinier than the more protected corresponding face, then cleavage may be suspected.

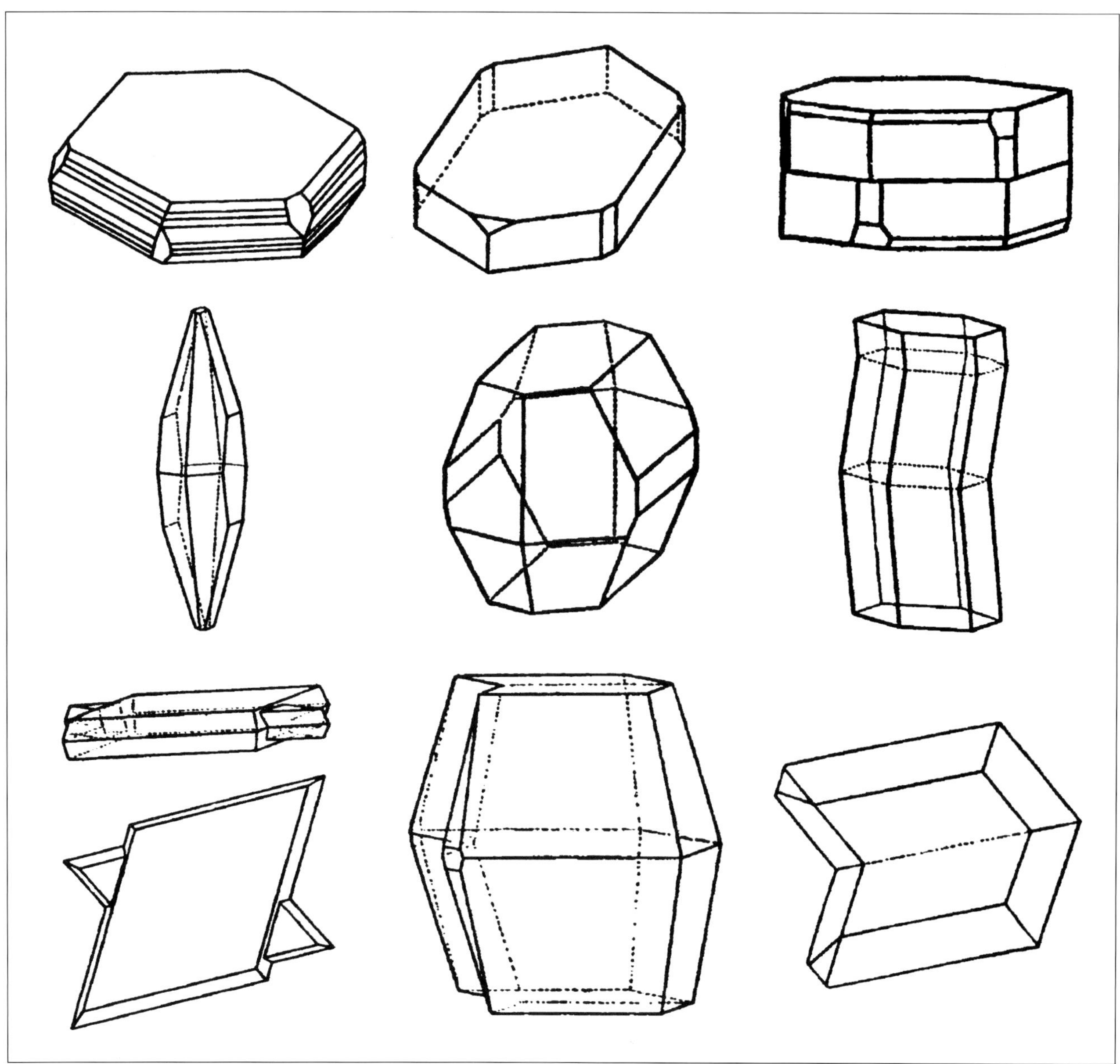

Figure 6. Drawings of some natural mica crystals, modified from Goldschmidt (1916). Some habits include tabular (top row), equant to elongated (middle row), and twinned crystals (bottom row).

Figure 7. A simple tabular crystal of phlogopite about 4 cm wide, with crude hexagonal outline, from Cantley, Quebec, Canada. RJL106

Figure 8. Elongated transparent golden yellow phlogopite crystal about 3 cm tall with tapering termination, in marble from Mogok, Myanmar. RJL2572

Figure 9. A twinned muscovite crystal about 3 cm across with a well-formed termination, from Lughman, Afghanistan. RJL2591

Micas provide excellent examples of twinning, perhaps the most distinctive of which are "fishtail" and "star" twins that can develop in muscovite crystals growing freely into a pegmatite cavity. Good specimens are found at Riacho Genipapo and elsewhere in Minas Gerais, Brazil, often associated with gem crystals and other pegmatite minerals.

Figure 10. A nice example of muscovite "fishtail twins" with garnet from Riacho Genipapo, Minas Gerais, Brazil. The garnet is likely a member of the spessartine-almandine series. RJL3213

In addition to the familiar thick to thin tabular habit, micas can equally well form elongated or columnar crystals. Aggregates of many columnar crystals growing in a subparallel or slightly divergent mass can give rise to interesting curved or cauliflower-like formations such as those found at Karabib, Namibia. Stresses during growth can also create extremely curved forms called "ball micas". Specimens of ball muscovites have been found at Fillow Quarry, Connecticut, and at several places in Maine, including: Lord Hill Quarry, Stoneham; Bennett Quarry, Buckfield; Mount Mica Quarry; Paris; and Maine Feldspar Quarry, Auburn (Vandall King, personal communication). Ball lepidolite is known from several locales in Minas Gerais, Brazil, including Parelhas and Aracuai; the largest ball micas are lepidolites from the Muiane Quarries, Alto Ligonha, Mozambique, where individual spheroids to 30 cm have been found.

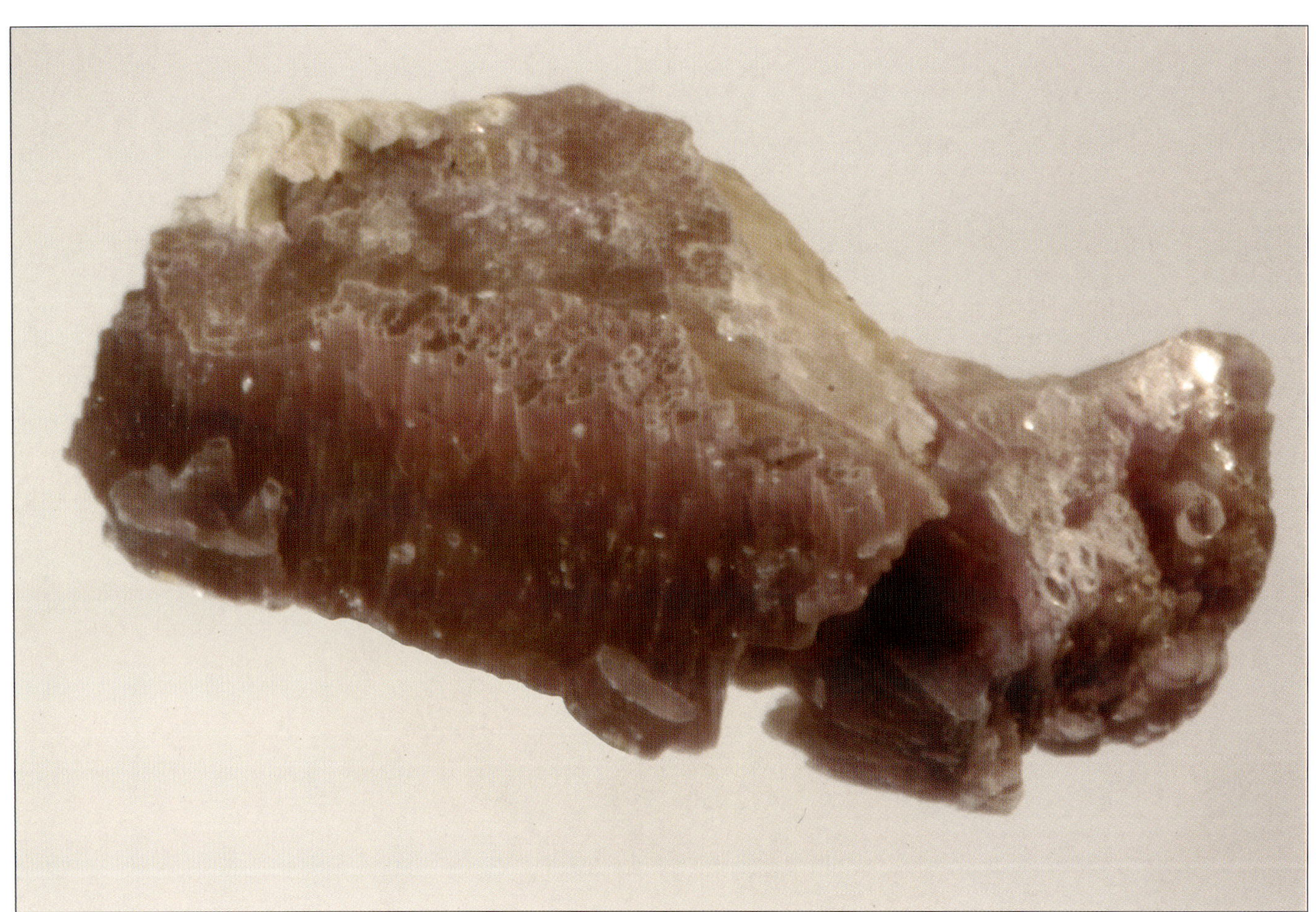

Figure 11. Pink lepidolite from the Himalaya mine, California, showing columnar crystals in parallel growth. Sample is about 5 cm wide. RJL818

Figure 12. Dark lavender lepidolite showing the rounded habit commonly called "ball mica." Sample is about 6 cm wide. RJL 3215

Figure 13. Small samples of muscovite "ball mica" from the Chandler mine, Raymond, New Hampshire.

Accepted species and their formulas

Table 1. True micas

Dioctahedral	
Muscovite	$KAl_2\square AlSi_3O_{10}(OH)_2$
Aluminoceladonite	$KAl(Mg,Fe^{2+})\square Si_4O_{10}(OH)_2$
Ferro-aluminoceladonite	$KAl(Fe^{2+},Mg)\square Si_4O_{10}(OH)_2$
Celadonite	$KFe^{3+}(Mg,Fe^{2+})\square Si_4O_{10}(OH)_2$
Ferroceladonite	$KFe^{3+}(Fe^{2+},Mg)\square Si_4O_{10}(OH)_2$
Chromceladonite	$KCr^{3+}(Mg)\square Si_4O_{10}(OH)_2$
Ganterite	$[Ba_{0.5}(Na,K)_{0.5}]Al_2(Si_{2.5}Al_{1.5}O_{10})(OH)_2$
Roscoelite	$KV_2\square AlSi_3O_{10}(OH)_2$
Chromphyllite	$KCr_2\square AlSi_3O_{10}(OH)_2$
Boromuscovite	$KAl_2\square BSi_3O_{10}(OH)_2$
Paragonite	$NaAl_2\square AlSi_3O_{10}(OH)_2$
Nanpingite	$CsAl_2\square AlSi_3O_{10}(OH)_2$
Tobelite	$(NH_4)Al_2\square AlSi_3O_{10}(OH)_2$
Trioctahedral	
Annite	$KFe^{2+}{}_3AlSi_3O_{10}(OH)_2$
Fluorannite	$KFe^{2+}{}_3AlSi_3O_{10}(F)_2$
Phlogopite	$KMg_3AlSi_3O_{10}(OH)_2$
Fluorophlogopite	$KMg_3AlSi_3O_{10}(F,OH)_2$
Siderophyllite	$KFe^{2+}{}_2AlAl_2Si_2O_{10}(OH)_2$
Eastonite	$KMg_2AlAl_2Si_2O_{10}(OH)_2$
Hendricksite	$KZn_3AlSi_3O_{10}(OH)_2$
Shirozulite	$KMn_3(SI_3Al)O_{10}(OH,F)_2$
Montdorite	$KFe^{2+}{}_{1.5}Mn^{2+}{}_{0.5}Mg_{0.5}\square_{0.5}Si_4O_{10}F_2$
Tainiolite	$KLiMg_2Si_4O_{10}F_2$
Polylithionite	$KLi_2AlSi_4O_{10}F_2$
Trilithionite	$KLi_{1.5}Al_{1.5}AlSi_3O_{10}F_2$
Shirokshinite	$K(NaMg_2)Si_4O_{10}F_2$
Masutomilite	$KLiAlMn^{2+}AlSi_3O_{10}F_2$
Norrishite	$KLiMn^{3+}{}_2Si_4O_{12}$
Tetra-ferri-annite	$KFe^{2+}{}_3Fe^{3+}Si_3O_{10}(OH)_2$
Tetra-ferriphlogopite	$KMg_3Fe^{3+}Si_3O_{10}(OH)_2$
Aspidolite	$NaMg_3AlSi_3O_{10}(OH)_2$
Preiswerkite	$NaMg_2AlAl_2Si_2O_{10}(OH)_2$
Ephesite	$NaLiAl_2Al_2Si_2O_{10}(OH)_2$

Table 2. Brittle micas

Dioctahedral	
Margarite	$CaAl_2\square Al_2Si_2O_{10}(OH)_2$
Chernykhite	$BaV_2\square Al_2Si_2O_{10}(OH)_2$
Trioctahedral	
Clintonite	$CaMg_2AlAl_3SiO_{10}(OH)_2$
Bityite	$CaLiAl_2BeAlSi_2O_{10}(OH)_2$
Anandite	$BaFe^{2+}{}_3Fe^{3+}Si_3O_{10}S(OH)$
Kinoshitalite	$BaMg_3Al_2Si_2O_{10}(OH)_2$
Ferrokinoshitalite	$BaFe^{2+}{}_3Al_2Si_2O_{10}(OH)_2$
Oxykinoshitalite	$Ba(Mg_2Ti)Al_2Si_2O_{10}(O_2)$

Table 3. Interlayer-deficient micas

Dioctahedral	Idealized formula: $(K,Na)_{x+y}(Mg,Fe^{2+})_x(Al,Fe^{3+})_{2-x}\square Si_{4-y}O_{10}(OH)_2$ where $0.6 \leq x+y < 0.85$; $Mg > Fe^{2+}$; $Al > Fe^{3+}$
Illite (series)	$K_{0.65}Al_2\square Al_{0.65}Si_{3.35}O_{10}(OH)_2$
Glauconite (series)	$K_{0.8}R^{3+}{}_{1.33}R^{2+}{}_{0.67}\square Al_{0.65}Si_{3.87}O_{10}(OH)_2$
Brammalite (series)	$Na_{0.65}Al_2\square Al_{0.65}Si_{3.35}O_{10}(OH)_2$
Trioctahedral	
Wonesite[a]	$Na_{0.5}\square_{0.5}Mg_{2.5}Al_{0.5}AlSi_3O_{10}(OH)_2$

[a] Wonesite is not an end member

Table 4. Series names

Biotite	**Trioctahedral micas between, or close to, compositions bounded by the lines between the end members annite and phlogopite and between siderophyllite and east-onite; dark micas without lithium**
Brammalite	**Dioctahedral interlayer-deficient micas with compositions defined in Table 3**
Glauconite	**Dioctahedral interlayer-deficient micas with compositions defined in Table 3**
Illite	**Dioctahedral interlayer-deficient micas with compositions defined in Table 3**
Lepidolite	**Trioctahedral micas on, or close to, compositions running between trilithionite and polylithionite; light-colored micas with substantial lithium**
Phengite	**Potassic dioctahedral micas between or close to compositions bounded by the lines running between muscovite and aluminoceladonite and between muscovite and celadonite**
Zinnwaldite	**Trioctahedral micas on, or close to, compositions running between siderophyllite and polylithionite; dark micas containing lithium**

Collectors will notice at once that many familiar names such as biotite, lepidolite, and zinnwaldite do not appear in Tables 1-3. These terms have been relegated to "series" status and individual members of the series properly have their own IMA-accepted species names as given in Table 4. This presents a dilemma for the collector who wants to have his specimens correctly labeled but doesn't have access to X-ray crystallography facilities. The only practical solution is to try to find literature or anecdotal analyses of material from the particular locale; failing that, it is best to use the series name (as we often do with "apophyllite" or "scapolite" specimens). For a competitive display, care must be taken to ensure that labeling is consistent with the local judging rules.

For the systematic collector, and particularly for the enthusiast who seeks to have at least one example of every valid mineral species, the mica group presents some interesting challenges and rewards. On the one hand most of the recognized species are probably available commercially in one form or another, so assembling a fairly complete collection of the species is not impossible or prohibitively expensive. On the other hand, many dealers will continue to use the series names, so it will be left to the collector either to determine the actual species through research, analysis, or an educated guess, or to simply retain the series name. For the aesthetic display collection, series names may be adequate (and will be used here in some cases). Collectors may take comfort in knowing that field geologists, who often need to log samples quickly without the benefit of advanced analytical instruments, frequently use "series" names because it is convenient and *it conveys the information that is truly needed to do the job*. This perspective is sometimes lost in purely academic discussions.

Figure 14. A thick crystal section of black "biotite" about 5 cm wide and 1 cm thick, from Topsham, Maine; specimen was obtained by the author c. 1971; under current nomenclature this material is annite. RJL153

As with many silicates, any particular mica species can display a significant range of compositions within the field assigned to that species and will be given its name based on the dominant ions on particular sites. Complete solid solution series extend between many of the compositions, and in some cases different zones in a single crystal might lie on opposite sides of a compositional boundary, in which case the crystal would actually represent an intimate mixture of two mineral species. A noteworthy example is the zinnwaldite found at Morefield mine, Virginia, where the analyses lie close to the divide between polylithionite and siderophyllite. An even more extreme case occurs in the rare-element pegmatites at Red Cross Lake, Manitoba, Canada (Hawthorne, Teertstra, and Černý 1999). There, five different micas were found with dominant Cs or Rb representing: Rb- and Cs- dominant polylithionite, Rb- and Cs- dominant magnesian annite, and Cs-dominant ferroan phlogopite. Each of these compositions could be a candidate for species status if enough material were present to allow a complete mineralogical characterization; however, each one was restricted to the margins of strongly zoned microcrystals and was insufficient for crystal structure analysis (Černý et al. 2003). In pegmatites of the Pikes Peak batholith, mica crystals were found with more than a hundred color zones in a single crystal; in this case the color zoning was correlated with oscillatory Ti concentration in the crystals (Foord et al. 1995). Zoned crystals from the Mokrusha pit, Alabashka pegmatite field, Middle Urals, Russia, show various combinations of species, for example a core of masutomilite and/or zinnwaldite surrounded by lepidolite and then a later rim of muscovite (Popova, Popov, and Kanonerov 2002).

Figure 15. A zoned crystal, about 4 cm across, whose composition ranges from a core of masutomilite (possibly also containing a zone of zinnwaldite) to a rim of lepidolite from the Mokrusha mine, Murzinka pegmatite district, Russia. RJL3204

Figure 16. A zoned crystal of transparent muscovite with a pale lavender rim of lepidolite, from Minas Gerais, Brazil. Crystal is about 5 X 9 cm and is backlit to highlight the zoning. RJL3214

Formation and Geochemistry

Micas in igneous and hydrothermal rocks

Micas that are believed to have crystallized from the melt, including the dioctahedral mica, muscovite, and trioctahedral micas of the biotite series, have contributed much to our understanding of igneous rocks. This includes such fundamental questions as their origin, emplacement, and the history and conditions of their crystallization. For example, the composition of the biotite in an igneous rock can be used as a tool to estimate either the maximum oxygen activity or temperature or the minimum water or hydrogen activity at which the biotite formed (Speer 1984). The fractionation of rare elements can be used to better understand the evolution and consolidation of complex pegmatites (Černý et al. 1995; Novák and Černý 1998).

For collectors, the finest specimens of muscovite and lepidolite are found in cavities in granite pegmatites, as is the rare species boromuscovite. Examples of this type of occurrence include: the gem pegmatites of San Diego County, California (Little Three, Pala, Himalaya, and others); pegmatite quarries of the Bumpus district and elsewhere in Maine; the Alabashka pegmatite field in Russia; and numerous localities in Minas Gerais, Brazil. Igneous occurrences of phlogopite of most interest to collectors are in carbonatites, exemplified by the calcite-apatite vein dikes in Haliburton County, Ontario, Canada, where excellent and sometimes huge crystals are found. Phlogopite is also found in ultrabasic rocks, particularly kimberlites where phlogopite can constitute as much as 20% of the rock.

Figure 17. A typical specimen of muscovite and albite from a cavity in pegmatite, showing the large, well-developed crystals. The largest muscovite crystal is about 7 cm across. Specimen is from Minas Gerais, Brazil. RJL1850

Figure 18. Crude crystals of muscovite about 1 to 2 cm across in albite from pegmatite at Smith Ledge, Auburn, Maine. RJL115

Figure 19. Earthy beige boromuscovite on lepidolite crystals about 2 cm across, from the Little Three mine, San Diego County, California. RJL3143

Some micas can be products of hydrothermal processes. Celadonites are found lining or completely filling vesicles in basalts (often associated with zeolites) or as replacements of ferromagnesian minerals including olivine and hypersthene; the fact that many celadonites are the 1M polytype suggests that they formed at temperatures typical of hydrothermal conditions (Odom 1984). Readers are cautioned that the foregoing statement applies specifically to celadonite and is not intended as a general rule; other micas may form the 1M polytype under other conditions. Roscoelite occurs in U- and As-Bi-Co-Ni veins at several locales in the Czech Republic. The fine ephesite crystals found at several mines in the Postmasburg district of South Africa were formed through metasomatic alteration of manganiferous sediments by hydrothermal fluids (Cairncross and Dixon 1995).

Figure 20. Brownish heulandite crystals in a pocket about 4 cm wide that is lined with earthy, blue-green celadonite. Sample is from Pato Branco, Parana, Brazil. RJL692

Figure 21. Bladed stilbite crystals colored dark green by included celadonite particles from Jalgaon, India. RJL2790

Micas in metamorphic rocks

Only a few mica species are common in metamorphic rocks and they show limited chemical variability; nevertheless it has been suggested that micas "may be the most useful group of minerals for gaining information on the petrogenesis of those metamorphic rocks that contain them" (Guidotti 1984). Micas can be used to define the limits of some metamorphic facies; e.g., the upper part of the amphibolite facies may be defined in terms of reactions related to the upper stability limit of muscovite, and the granulite facies can be defined by the decomposition of biotite to sillimanite + hypersthene.

Muscovite is commonly present in regionally metamorphosed sediments and can be found in each of the progressive stages or "zones" of metamorphism. In the earliest stage, or "chlorite zone," muscovite forms from the recrystallization of claylike particles in the sediment, and the appearance of muscovite in argillaceous rocks is often the first clear sign of metamorphism (Deer, Howie, and Zussman 1962).

Phlogopite is a common constituent of marbles where it sometimes forms spectacularly large, transparent crystals. In the French Pyrenees, for example, it is found in contact metamorphosed dolomite marbles adjacent to the Querigut granite. In some locales, it is the product of thermal metasomatism in which fluorine was introduced from F-rich pegmatite intrusives. Phlogopite also forms during regional metamorphism of impure magnesian limestones. When excess alumina is present, dolomite + muscovite can react to form phlogopite + corundum or phlogopite + spinel. This process can yield fascinating specimens of golden-brown to greenish phlogopite with colorful, gemmy corundum or spinel crystals.

Figure 22. Large transparent brown phlogopite crystals (largest is about 3 cm tall) in white marble from Mogok, Myanmar. RJL2290

Figure 23. Golden tabular phlogopite with purple spinel from Myanmar, demonstrating the reaction dolomite + muscovite = phlogopite + spinel that occurred during formation of the metamorphic rock. This sample was undoubtedly formed in a matrix of dolomitic marble that was removed by acid etching. RJL2932

Figure 24. An iron-stained sample of mica schist from Freeport, Maine, in which the mica is the green chromian variety of muscovite (*fuchsite*). Sample is about 5 cm wide and the individual muscovite flakes are about 1 mm wide. RJL397

The brittle micas are characteristically found in metamorphic rocks. Margarite occurs in metamorphic emery deposits associated with diaspore and corundum; it is also found in chlorite and mica schists associated with tourmaline and staurolite. Clintonite is commonly found in metasomatically altered limestones associated with spinel, grossular, vesuvianite, and phlogopite; it is also found in some skarns at contacts between dolomitic marbles and granite (Deer, Howie, and Zussman 1962).

Figure 25. A hand specimen of clintonite in blue marble from Crestmore, California. Individual clintonite crystals are about 3 mm across. RJL739

Micas in sedimentary rocks

Many sedimentary rocks contain micas, although they are rarely of interest to the collector because they are, for the most part, microscopic clay-sized particles of illite. Perhaps the most interesting example of mica that is characteristically found in sedimentary environments is glauconite. The mineral typically forms tiny (sub-millimeter) pellets that are found extensively in sedimentary rocks of widely different ages from Cambrian to Recent. Glauconite is also found in Recent marine sediments on the continental shelves bordering most of the major continents and Japan. Glauconites are chemically similar to the celadonites except that glauconites tend to form in sedimentary (typically marine) environments, whereas celadonites tend to be associated with igneous rocks, usually basalts, where they are found as vesicle fillings or occasionally replacing olivine or hypersthene (Odom 1984).

The vanadium mica, roscoelite, is found as a cementing phase in sandstone at several locales, including the Mounana mine, Gabon, and in the Colorado Plateau region.

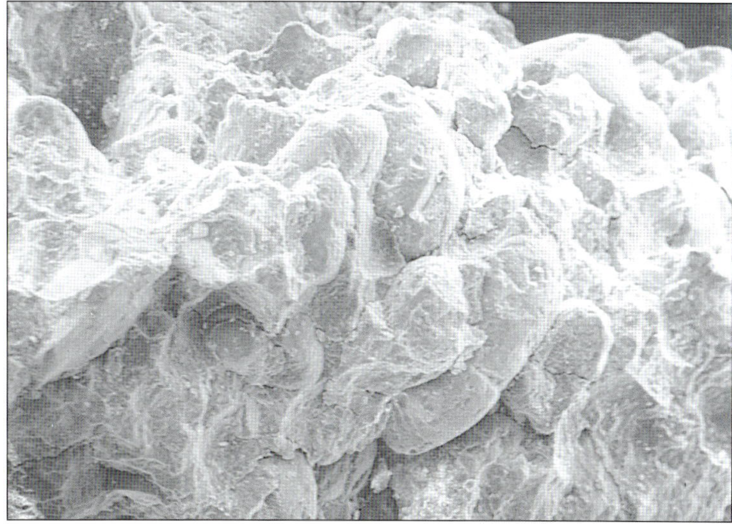

Figure 26. Scanning electron micrograph of glauconite from Oued Kasseb, Tunisia, showing the characteristic spheroidal particles. Field of view is about 2 mm wide. RJL731

Figure 27. Photomicrograph of fine-grained sandstone from Placerville, Colorado, with minute silvery flakes of roscoelite distributed between rounded quartz grains. RJL723

Micas in meteorites

Minerals of the mica group have been found in a number of meteorites and occasionally in interplanetary dust particles, making them of interest to planetary researchers who seek to gain insights into conditions on other planets. For example, in the Allende CV3 carbonaceous chondrite, clintonite replaces grossular in alteration veins and anorthite contains lamellae of margarite. Because chondrite-type meteorites are not fragments of planetary bodies, the presence of hydrous phases is attributed to late hydrous alteration in the solar nebula. The Lujiang, China stony meteorite, which was collected as a fall, contains both muscovite and phlogopite. Phlogopite has been found in melt inclusions in the Chassigny meteorite, which is thought to be a fragment from the surface of Mars. Another putative Martian meteorite, Allan Hills 84001, contains potassium mica. Both of these meteorites are of particular interest for establishing the existence of hydrous minerals on planets other than Earth (Fleet 2003).

Figure 28. A slice of the Allende carbonaceous chondrite, one of the meteorites in which mica has been observed. Pale chondrules to about 2 mm are distributed in a dark groundmass. RJL1177

The Minerals

Aluminoceladonite was originally described in 1883 from Wiesmath and Frohsdorf, Lower Austria as "leucophyllite." That name has been rejected because it is also the name of a rock type; furthermore, the original analysis is suspect, so for some time aluminoceladonite was viewed as a hypothetical end member that might or might not have been found in nature, particularly at the purported type locale. Within the last few years, aluminoceladonite has been reported from several locales, including: the Dora Maira Massif, Piedmont, and the Praborna mine, Saint-Marcel, Val d'Aosta, Italy; Narayama, Niigata Pref., Japan; and Podsulova, Rudohorie Mts., Slovak Republic. Known polytypes: $2M_1$, 1Md

Anandite is an unusual sulfur-bearing brittle mica (Filut et al. 1985) that forms black, nearly opaque veins and lenses in a banded magnetite deposit at the Wilagedera prospect, North Western Province, Sri Lanka (type locale, TL). At the Esquire No. 7 claim, Big Creek, Fresno County, California, it is found with a complex suite of barium-containing species in a quartz-sanbornite rock. It has also been reported from barium silicate rich lenses in quartzite at Trumbull Peak, Mariposa County, California; and from several sites in the Franklin mining district, Sussex County, New Jersey. Known polytypes: **2Or**, $2M_1$, 1M.

Figure 29. A thumbnail sized specimen containing silvery flakes of aluminoceladonite from the Dora Maira Massif, Cuneo Province, Piedmont, Italy. RJL3040

Figure 30. Black micaceous anandite with bright red gillespite (a rare barium silicate) from Fresno Co., California. RJL3190

Figure 31. Photomicrograph of black platy crystals of anandite from Fresno Co., California. Field of view is about 5 mm wide. RJL890

Annite forms tabular crystals and thin cleavage fragments with pseudohexagonal outline, or foliated masses. It is abundant in granite pegmatites around the world, and most "biotite" lies within the annite compositional field. It is named for the type locale, Cape Ann, Massachusetts. Other localities include: Pikes Peak, Colorado; Mineral Co., Nevada; Wilberforce, Ontario, Canada; Mont Saint-Hilaire, Quebec, Canada; the Moina skarn, Tasmania, Australia; the Kawai mine, Ena, Gifu Pref., Japan; Flowerdale, Scotland; in Russia at Katugin, Siberia and at Sludorudnik, Ural Mountains. Known polytype: 1M.

Aspidolite, first described in 1869, has more recently been called "sodium phlogopite" by some authors; however, the name aspidolite is recommended (Rieder et al. 1998). At Derag, Tell Atlas, Algeria this species forms minute flakes, rimmed by normal phlogopite, in a massive dolomite in a metamorphic evaporite deposit. Known polytype: 1M.

Figure 32. Black annite crystals to about 15 mm on matrix from Garfield Hills, Mineral County, Nevada. RJL 2905

Figure 33. Black micaceous masses of annite about 1 X 2 cm in matrix, from Katugin River, Siberia, Russia. RJL3189

Bityite is a fairly rare brittle mica having Be occupying some of the tetrahedral sites. It forms small, thin tabular crystals (2 mm) but more commonly rosettes and crusts of tiny flakes, in lithium pegmatites. Individual crystals are transparent, yellowish to pale brown, and display pale yellow fluorescence under SW UV light. At Maharitra, Mt. Bity, Madagascar (TL) it is associated with albite, lepidolite, and pink tourmaline; at Londonderry, Western Australia it occurs with albite, beryl, bavenite, columbite, and cassiterite. It is also found at the Foote mine, Kings Mt., North Carolina; at the Harding pegmatite, Taos County, New Mexico; and at Val Vigezza, Piemont, Italy. Known polytype: $2M_1$.

Boromuscovite was first described as "boron-substituted muscovite" from the New Spaulding pocket at the Little Three mine, San Diego County, California (Foord et al. 1989). It was later given formal species status (Rieder et al., 1998). At the type locale, it forms white to off-white, porcelaneous coatings similar to the pocket clays found at the nearby Himalaya dike system, but somewhat harder. Lavender crystals of lepidolite dusted with white boromuscovite make particularly attractive specimens. Similar material has been found at the Mika pegmatite, Tadjikistan. Boromuscovite has also been reported from an elbaite pegmatite at Rečice, Moravia, Czech Republic (Liang et al. 1995; Novák et al. 1999). Known polytypes: $\mathbf{2M_1}$, 1M (type material is a 50:50 mixture of the two, whereas at Rečice the ratio is about 83:17).

Brammalite refers to compositions that might be thought of as the sodium analogue of illite. It forms white to pale green scales and fibrous to earthy masses. It is often found in shales or mudstones associated with coal beds, e.g., at Llandybie, Dyfed, Wales (TL); Pilot Knob, Ironton, Missouri; and the Pittsburgh coal seam, Pursglove, West Virginia. Some occurrences are believed to have resulted from the degradation of paragonite (Gaines et al. 1997). Known polytype: $2M_1$.

Figure 34. Minute tan masses of bityite in white marble, from Val Vigezzo, Italy. Field of view is about 5 mm across. RJL3202

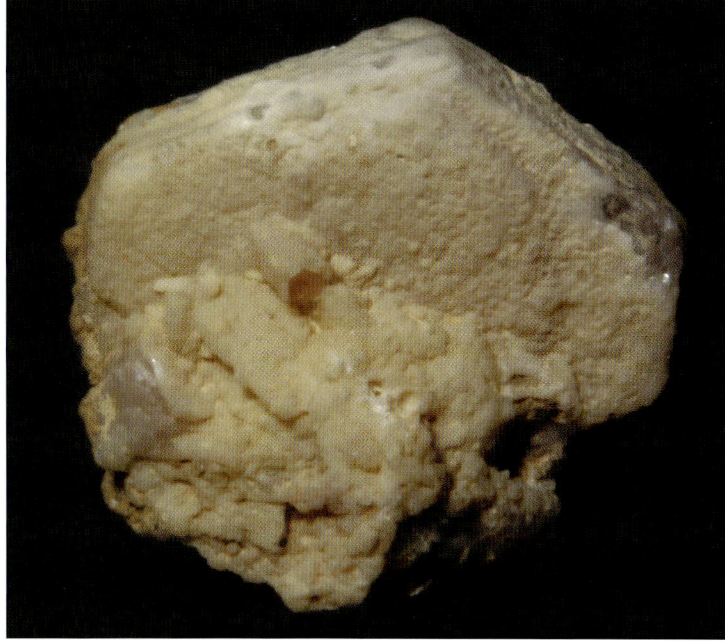

Figure 35. Beige porcelaneous coating of boromuscovite on a tabular lepidolite crystal about 2 cm across, from the Mika pegmatite, Turakuloma Range, Tadjikistan. RJL2844

Celadonite forms sea-green earthy aggregates and coatings, typically as vesicle linings and coatings in altered volcanic bodies of intermediate to basaltic composition. It is the green phase seen on the familiar "amethyst pipes" of Rio Grande Sul, Brazil and on many zeolite specimens. Spectacular, opaque dark green fluorapophyllite crystals from Jalgaon, India are colored in part by celadonite inclusions. This mineral and the other members of the series are virtually indistinguishable from one another and occasionally form intergrown mixtures. Known polytype: 1M.

Chernykhite, a brittle mica that represents the Ba analogue of roscoelite, is olive green with a pearly luster and typical micaceous cleavage. It occurs at Kara Tau, Zhambil, Kazakhstan, where it forms veins in carbonate rocks (Ankinovich et al. 1973). Known polytype: $2M_1$.

Figure 36. Large crystals of apophyllite colored dark green by celadonite inclusions, from Jalgaon, India. RJL2343

Figure 37. A transparent "amethyst flower" about 9 X 12 cm, from Rio Grande do Sul, Brazil; green color comes from the underlying layer of celadonite. RJL 2980

Figure 38. Detailed view of the rear of the specimen shown in Figure 37, showing celadonite-lined vesicles. RJL2980

Chromceladonite was recently described from the Srednyaya Padma uranium-vanadium deposit, South Karelia, Russia (Pekov et al. 2001), where it occurs as aggregates of thin, green to dark green lamellae to 1 cm, and as veinlets and spherulites. It appears to be of metasomatic origin. Roscoelite and chromphyllite are also reported from this locale, along with calcite, hematite, uraninite, zincochromite, and vanadium oxides and selenides. Known polytype: 1M.

Figure 39. Photomicrograph of dark green chromceladonite from Srednyaya Padma, Russia. RJL3207

Chromphyllite is the chromium analogue of muscovite. It forms emerald green, 0.3 – 0.4 mm plates and flakes, often containing inclusions of chromite and eskolaite, in the parametamorphic Slyudyanka complex, Baikal region, Siberia, Russia, in Cr-enriched layers in quartz-diopside rocks. Associates include muscovite, phlogopite, uvarovite, chromian dravite, and other Cr minerals (Reznitsky et al. 1998). In the Outokumpu district, Finland, it occurs in quartzite, associated with uvarovite, staurolite, diopside, tremolite, epidote, and chlorite. The formation of chromphyllite is attributed to local Cr metasomatism under low- to medium-grade metamorphic conditions, with local ultramafic rocks or detrital chromite serving as the source of Cr (Fleet 2003). Known polytype: $2M_1$.

Figure 40. Pale gray quartzite with bands of green chromphyllite, from the type locale, Slyudyanka, Russia. RJL3039

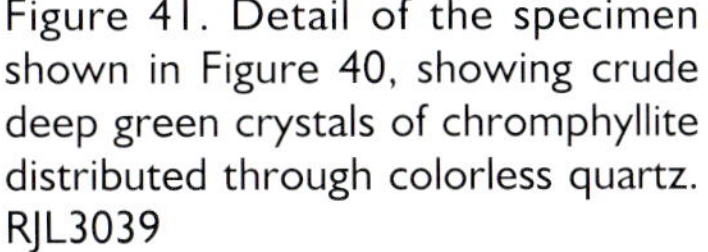

Figure 41. Detail of the specimen shown in Figure 40, showing crude deep green crystals of chromphyllite distributed through colorless quartz. RJL3039

Clintonite occurs mainly in crystalline limestones in association with calcite, grossular, vesuvianite, clinopyroxene, phlogopite, and spinel. It is also sometimes found in chloritic schists and contact zones. Tabular greenish crystals of clintonite var. xanthophyllite, in blue calcite from the Crestmore quarry, Riverside County, California, are typical of the species; a recent find of crystals up to about 3 cm at Crestmore set a new standard of quality for this mineral. Known polytypes: **1M**, $2M_1$, 3T.

Figure 42. Simple tabular green clintonite crystals about 3 mm wide in blue marble from Crestmore, California. RJL739

Figure 43. Dark green clintonite crystals to about 1 cm wide, from the recent find at Crestmore, California. RJL3155

Eastonite is considered by some authors to be an aluminous variety of phlogopite, but it has been given species status (Rieder et al. 1988). The type locale is the Williams quarry, Easton, Pennsylvania; it is known from several other Pennsylvania localities in Northampton County (all near Easton), and the material has been variously described in the literature as yellow green to greenish gray to olive green. Virtually all of the material is some greenish shade, ranging from silvery green to darker olive green. It is pearly to vitreous, flexible but inelastic, and scaly. The old descriptions of "chloritic vermiculite" appear to describe similar compositions, high in Mg content (30 wt% or more in most cases) (A. J. Nikischer, personal communication). Eastonite has also been reported from Liset, Sogn og Fjordane, Norway. Known polytype: 1M.

Figure 44. Photomicrograph showing greasy blackish-green masses of eastonite forming pods in quartz from the type locale, Easton, Pennsylvania. RJL3208

Ephesite was first described in 1851 from the emery deposit at Gumuchdagh, Ephesus, Turkey. It has also been reported from the Ilimaussaq alkaline complex in southern Greenland. It forms pink to pinkish brown crystals, often with a pearly luster on the basal (cleavage) plane. More recently, excellent specimens have been produced from the Lohatla, Bishop, and Gloucester mines in the Postmasburg manganese fields, Cape Province, South Africa, where deep pink to red crystals up to several cm wide are associated with bixbyite, manganoan diaspore, pyrolusite, manganite, and several other manganese minerals (Cairncross and Dixon 1995). Known polytypes: **$2M_1$**, 2A, 1M.

Ferro-aluminoceladonite occurs in an altered tuff, intergrown with ferroceladonite as a dark blue-green earthy mixture, in the Hokonui Hills, Southland, New Zealand (Li et al. 1997). It has also been reported from Tai-Keu, Polar Urals, Russia. Known polytype: 1M.

Figure 45. Pearly pink ephesite crystals to several mm across, forming a solid mass, from Gloucester mine, South Africa. RJL725

Figure 46. Photomicrograph showing tiny grains of dark green ferro-aluminoceladonite, from Tai-Keu, Polar Urals, Russia. RJL3038

Ferroceladonite occurs with ferro-aluminoceladonite at the type locale noted above (Li et al. 1997). At the Poudrette quarry, Mont Saint-Hilaire, Quebec, Canada, green microcrystals of ferroceladonite with other minerals such as elpidite, epididymite, microcline, and monazite may be of some interest to micromount collectors. It has also been found at Mikhailovskoe, Russia. Known polytype: 1M.

Ferrokinoshitalite, the Fe^{2+} analogue of kinoshitalite, occurs in a banded iron formation at the Broken Hill mine near Aggeneys, northern Cape Province, South Africa, associated with quartz, magnetite, spessartine, and several Mn-rich silicates. It forms minute (0.2 mm) dark green tabular crystals (Guggenheim and Frimmel 1999). Known polytype: 1M.

Figure 47. Photomicrograph of deep green ferroceladonite forming rich schistose masses, from Mikhailovskoe, Russia. RJL3037

Fluorannite, the F-dominant analog of annite, was described from an A-type granite at Suzhou, near Shanghai, China (Shen, Lu, and Xu 2001). It forms black, submetallic euhedral to subhedral tabular crystals and sheets up to 6 mm wide. It has also been reported from the Katugin Ta-Nb deposit, Chitinskaya Oblast, Transbaikalia, Russia. Known polytype: 1M.

Fluorophlogopite, the F-dominant analogue of phlogopite, was first described from Mt. Calvario, Biancavilla, Etna Volcanic Complex, Sicily, Italy. At the type locale, the mineral is found in autoclasts of gray-red altered lava, associated with fluoro-edenite, alkali feldspars, clino- and orthopyroxenes, fluorapatite, hematite, and pseudobrookite. Its formation is attributed to metasomatic alteration of the original lava by a very hot fluid enriched in F and Cl. Crystals are pale yellow, forming thin laminae about 200 to 400 μm in diameter (Gianfagna et al. 2007). At Ristiniemi Cape, Pitkyaranta, Karelia, Russia, small gray-brown crystals are found in white marble, along with tiny orange chondrodite grains. It has also been reported from: Roc de Courlande, Puy-de-Dome, France; Hillesheim, Eifel Mts., Rhineland-Palitanate, Germany; Uroi Hill, Romaina; and the St. Joe No. 3 mine, Balmat, New York.

Ganterite, the Ba-dominant analog of muscovite, was first described from the crystalline basement rocks of the Berisal Complex, Simplon Region, Switzerland, where it forms light gray to silver crystals (typically < 0.5 mm) in small bands or lenses 0.5 to 10 cm thick. It is interesting to note that although Ba^{2+} is the dominant interlayer cation, both Na^+ and K^+ are present and (Na + K) > Ba. Thus, because the divalent ion comprises less than half the total interlayer ions, ganterite is a true mica and not a brittle mica (Graeser, Hetherington, and Giere 2003). It has also been reported from the Lincoln Hill dumortierite deposit near Oreana, Nevada (Ma and Rossman 2006). Known polytype: $2M_1$.

Figure 48. Platy black crystals of fluorannite about 2 to 3 mm wide in matrix from Katugin, Russia. RJL3174

Figure 49. Small gray-brown tabular fluorophlogopite crystals in white marble, associated with tiny orange pods of chondrodite, from Ristiniemi Cape, Pitkyaranta, Karelia, Russia. RJL3126

Glauconite as presently defined comprises a series of compositions of generally green, clay-like materials that are similar to celadonite but differ in their mode of occurrence. Glauconite typically forms in marine environments, and glauconitic clays are of widespread occurrence; the mineral is actively being formed along continental margins. Several compositional variants have been reported from various localities (Gaines et al. 1997) and it is possible that further study will define a series of minerals analogous to the celadonites. Known polytypes: **1M**, **1Md**.

Figure 50. Dark greenish gray mass of glauconite, from Oued Kasseb, Tunisia. RJL731

Hendricksite is the Zn analogue of phlogopite; it has been reported only from the Franklin, New Jersey, mining district (Frondel and Ito 1966) but is fairly widespread there. It forms blackish to coppery brown euhedral crystals up to about 15 cm as well as larger plates in intergrown or deformed masses. Hendricksite is found in irregular lenses and sheets of skarn in a metamorphosed stratiform zinc deposit, associated with andradite, rhodonite, franklinite, willemite, calcite, axinite and other species. Known polytypes: **1M**, $2M_1$, 3T.

Figure 51. Dark reddish black crystalline masses of hendricksite in matrix, from Franklin, New Jersey. RJL728

Illite refers to a series of clay minerals that are of worldwide distribution. These interlayer cation deficient micas are common in sediments, clays, marls, shales, and some slates (Gaines et al. 1997). Although illite never forms macroscopic crystals, it sometimes replaces crystals of other species, making it of some interest to pseudomorph enthusiasts. For example, some analyses of the sharp "pseudoleucite" crystals from Kershehir, Turkey indicate the presence of illite. Known polytypes: **1M**, **1Md**, $2M_1$, 3T.

Kinoshitalite forms tiny (< 1 mm) yellow-brown scales in the Misago orebody, Noda-Tamagawa mine, Iwate Pref., Honshu, Japan in a hausmannite-tephroite ore (Yoshii et al. 1973). It has also been reported from Hokkejino, Kyoto Pref., Japan; and Trumbell Peak, Mariposa County, California. Kinoshitalite also occurs in Mn-rich rocks intruded by silicic pegmatite and carbonate veins at Netra, Balghat district, Madhya Pradesh, India (Dasgupta et al. 1989). Known polytypes: **1M**, $\mathbf{2M_1}$.

Figure 52. Slab of clay-like illite about 4 cm wide, forming casts after halite crystals, from Bohling Farm, Nebraska. RJL3201

Figure 53. Photomicrograph showing silvery flakes of kinoshitalite in matrix, from Kyoto, Japan. RJL483

Lepidolite was described by Klaproth in 1792 from the Rozna pegmatite in western Moravia, Czech Republic (Černý et al. 1995) and under current nomenclature the name applies to a series rather than a single species; however, collectors will often encounter mineral specimens labeled "lepidolite." The material may be colorless to lavender-pink, and generally forms in granitic pegmatite cavities. The material is often fine-grained or scaly, but well-formed crystals are known from a number of localities. At the Stewart pegmatite at Pala, California, compact lepidolite masses containing small, elongated prismatic crystals of red elbaite are well known to collectors.

Figure 54. Interesting cluster of lepidolite from Virgem da Lapa, Minas Gerais, Brazil: several thick crystals to about 2 cm wide are overgrown by smaller columnar to drusy crystals. RJL690

Figure 55. Radiating "books" of pale pinkish gray lepidolite crystals on albite with orange spessartine; a characteristic association from Little Three mine, San Diego County, California. RJL2296

Figure 56. Thumbnail-sized mass of gemmy pink lepidolite, from the Himalaya mine, San Diego County, California. RJL744

Freestanding lepidolite crystals and divergent "books" are found at the Little Three mine near Ramona, California, associated with orange spessartine and white albite (Foord et al. 1989). Large thick translucent lavender crystals are found at Aracuai, Minas Gerais, Brazil. Distinctive spherical and stalactitic masses are once again being recovered from the Zambesi region of Mozambique. Sharp lavender crystals are also found in the Varutrask pegmatite, Sweden. Known polytypes: $\mathbf{2M_2}$, **1M**, 3T, $2M_1$.

Figure 57. Crude purple lepidolite crystal about 2 cm wide, in matrix, from the Varuträsk rare-element pegmatite, Sweden. RJL1938

Figure 58. Lavender-pink lepidolite from Rumford, Maine. RJL114

Figure 59. A thick single crystal of lepidolite, about 3 X 4 X 11 cm, from Aracuai, Brazil. Lighting at right edge emphasizes the translucency of this large crystal. RJL2395

Figure 60. Columnar lepidolite from Karabib, Namibia, forming a cauliflower-like mass. RJL2991

Margarite rarely forms well-defined tabular crystals. The usual habit consists of random aggregates of thin, pearly flakes, which are typically pink to grayish pink, pale yellow, or pale green. Margarite is found in low- to medium-grade metamorphic rocks such as chlorite and mica schists, glaucophane-bearing rocks, and emery deposits. Noteworthy localities include: Mt. Greiner, Tirol, Austria; the Chester Emery mines, Hampden County, Massachusetts; Corundum Hill, Chester County, Pennsylvania; near Meadow Valley, Plumas County, California; Mt. Yatyrgvata, northern Caucasus, Russia; and the Shin-Kiura mine, Oita Pref., Japan. Chromian margarite occurs at Line pit, State Line district, Maryland. Margarite pseudomorphs after kyanite occur at Glen Esk, Scotland. White to colorless margarite formed as an alteration of andalusite var. chiastolite is found in the Smalls Falls area, Maine. Known polytypes: **$2M_1$**, 1Md, 1M.

Figure 61. Foliated masses of pearly pink margarite from the Chester emery mine, Massachusetts. RJL733

Figure 62. Pearly greenish flakes and masses of margarite from Franklin, New Jersey. RJL732

Figure 63. Photomicrograph of specimen in the previous figure, showing small green masses of margarite. RJL732

Masutomilite is pale purplish pink with a vitreous to pearly luster. At the type locale in the Tanaka-miyama district, Shiga Pref., Honshu, Japan it forms druses and crystals up to 3 cm across and 1 cm thick associated with topaz, schorl, albite, and quartz (Harada et al. 1977). Elsewhere in Japan it has been reported from Tawara, Gifu Pref. It also occurs in lithium pegmatite at Ctidružice, Czech Republic as well as with zinnwaldite in miarolytic cavities in the Sawtooth batholith, Idaho. Fine crystals found in the Mokrusha pegmatite vein, Alabashka pegmatite field, Russia are typically zoned with areas of zinnwaldite, masutomilite, and lepidolite in the same crystal (Popova, Popov, and Kanonerov 2002). Interestingly, similar zoning was reported in the type material from Japan. Known polytypes: **1M** , $2M_1$.

Montdorite forms minute (5 to 25 μm) green to brownish-green grains in peralkaline rhyolite at the Mt. Dore stratovolcano, near La Bourboule, Puy-de-Dome, France (Gaines et al. 1997). Montdorite represents a rare class of Al-poor or "tetrasilicic" micas that had been known as synthetic phases since the mid-1960s but not found in nature until 1979. Known polytypes: **1M**, 3T(?).

Figure 64. Small iron-stained cleavage of transparent pale pink/violet masutomilite from the Hisu Kawa mine, Gifu Prefecture, Japan. RJL3203

Muscovite is widespread and abundant in many rock types, but from the collector's viewpoint well-crystallized material is nearly always found in granite pegmatites, from miarolytic cavities in granites, or from cavities in some quartz veins (Sinkankas 1964). It is usually nearly colorless, silvery gray, or pale tan, but may be green, pink, yellow, or brown. The interesting twinned variety, *star mica*, occurs in floater groups with quartz at Riacho Genipapo, near Aracuai, Minas Gerais, Brazil (Cassedanne and Cassedanne 1981). A spectacular recent find of muscovite was at the Jose Pinto mine, where over a ton of beautifully crystallized specimens were recovered, many associated with apatite and other species (Cassedanne and Alves 1990). Deep pink (manganoan) varieties are found at the Harding mine, Taos County, New Mexico (Jahns and Ewing 1977). *Fuchsite* (pronounced "FEWKS-ite") is a green chromian variety found in massive or schistose form at several localities, including Freeport, Maine, and on the Kola Peninsula, Russia (where it is associated with kyanite crystals). Well-formed silvery-green crystals of fuchsite occur at Karelia, Russia. *Alurgite* is a red variety containing Mn. Known polytypes: $\mathbf{2M_1}$, 1M, 1Md, 3T, $2M_2$.

Figure 65. Bright golden muscovite crystals (largest is about 25 mm wide) on albite from Galileia, Minas Gerais, Brazil. RJL 2295

Nanpingite is the cesium analogue of muscovite. It forms colorless to white scales and plates up to 10 mm across in radiating aggregates and occasionally as well formed crystals in the pollucite-rich middle zone of a muscovite-albite-spodumene pegmatite in the Nanping mine, Fujian Province, China associated with montebrasite, quartz, and apatite. Its formation is attributed to late-stage evolution of a large pegamatite vein that crystallized from residual hydrothermal fluids that had very high Cs content (Yang et al. 1990). The species is of some scientific interest because it crystallizes in the $2M_2$ polytype, which is rare among dioctahedral micas; this is attributed to the large size of the Cs^+ interlayer ion (Ni and Hughes 1996). Nanpingite has also been reported from granitic pegmatites of the Stargazer claim, O'Grady Lake area, Northwest Territories, Canada. Known polytype: $2M_2$.

Figure 66. A typical example of the twinned muscovite referred to as "star mica," from Minas Gerais, Brazil. RJL2447

Figure 67. A fine example of the large muscovite crystals recovered in huge quantities from the Jose Pinto mine, Minas Gerais, Brazil. RJL2104

Figure 68. Pale brown muscovite crystals on white albite, from Hickory, North Carolina. RJL2677

Figure 69. Silvery green crystals of muscovite var. *fuchsite* from Karelia, Russia. RJL1965

Figure 70. A large mass of deep red muscovite var. *alurgite* from Ouro Preto, Minas Gerais, Brazil. RJL3086

Figure 71. Bright red muscovite var. *alurgite* forming rich veins and layers in the matrix, from San Marcel, Val d'Aosta, Italy. RJL3104

Figure 72. A small cluster of gemmy, ric lexico. RJL 405

Norrishite is found as mm-sized, shiny, brown-black crystals in significant amounts on the dumps of the abandoned Hoskins mine, west of Grenfell, New South Wales, Australia, associated with pectolite-serandite, manganoan alkali pyroxene, braunite, and carbonates (Eggleton and Ashley 1989). It has recently been reported from the Wessels mine, South Africa, where it forms cm-sized black aggregates in dark purple sugilite (Gnos, Armbruster, and Villa 2003). Norrishite has the interesting distinction that its water content is the lowest of all analyzed micas. Known polytype: 1M.

Oxykinoshitalite, the oxy-analog of kinoshitalite, was recently described from an olivine nephelinite on Fernando de Noronha Island, Pernambuco, Brazil. It is bright orange to brown and strongly pleochroic. At the type locale it is associated with olivine, clinopyroxene, Fe-Ti oxide, nepheline, calcite, apatite, and K-rich feldspar (Kogarko et al. 2005). Known polytypes: not reported.

Figure 73. Tiny dark plates of norrishite on matrix, from the Cerchiara mine, La Spezia, Liguria, Italy. Field of view is about 5 mm wide. RJL3177

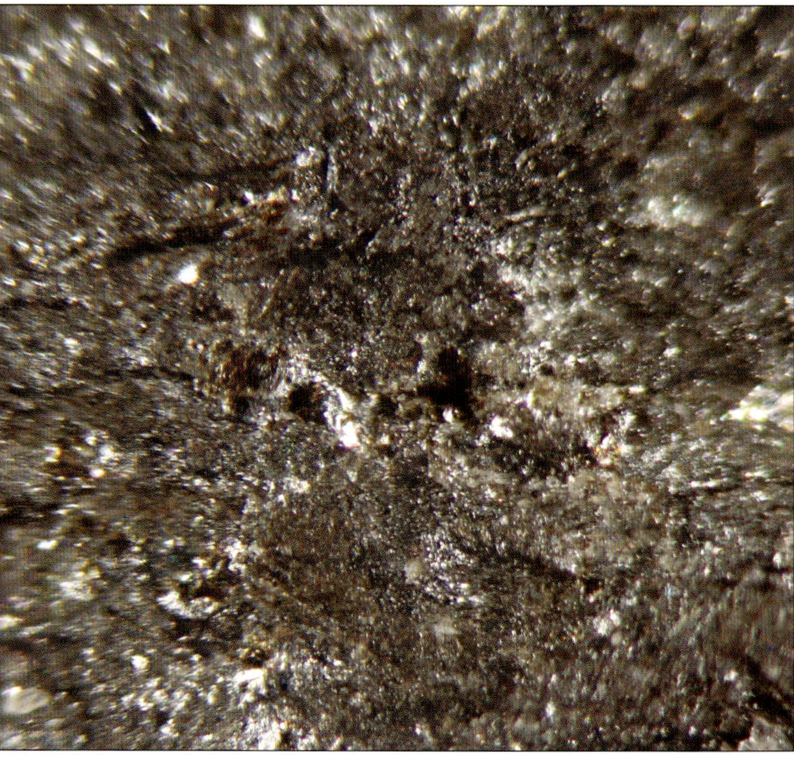

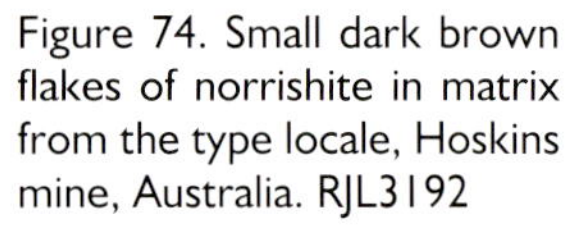

Figure 74. Small dark brown flakes of norrishite in matrix from the type locale, Hoskins mine, Australia. RJL3192

Paragonite is usually found in massive to scaly, fine-grained aggregates that might be mistaken for talc or muscovite. Before the use of X-ray diffraction and electron microprobe analysis, paragonite was considered to be a fairly rare mineral and many details of its chemistry were lacking. In the last thirty years this situation has improved greatly, and a better understanding has emerged on paragonite and the nature of its occurrences (Guidotti 1984). It has been reported from well over 100 localities worldwide, and is probably much more widespread than in just the well-documented occurrences. It is found in low- to medium-grade metamorphic schists and phyllites, in muscovite–biotite gneisses, quartz veins, fine-grained sediments, and glaucophane-bearing rocks. Some well-characterized occurrences include: Valais, Switzerland; Pragraten, Tirol, Austria; Ochsenkopf, Schwarztenberg, Saxony, Germany; Ivigtut, Greenland; Corundum Hill, Chester County, Pennsylvania; and the Leadville district, Lake County, Colorado. Known polytypes: $\mathbf{2M_1}$, 3T, 1M.

Figure 75. Silvery paragonite in schist, from Gassetts, Vermont. RJL484

Figure 76. Pearly gray flakes of paragonite distributed through matrix, from Oya Cho, Japan. RJL3209

Phlogopite is commonly found in metamorphosed limestone, where it typically forms sharp elongated or tabular crystals. Gemmy brown transparent crystals in either habit are found at Badakhshan, Afghanistan; similar transparent pale green to tan crystals occur in white marble at Mogok, Myanmar (Burma). It is widespread in the Grenville metamorphic limestones and calcite vein dikes of eastern Canada, where occasionally enormous crystals have been found. Large columnar crystals to 30 cm have been found at Franklin, New Jersey. Dark brown tabular crystals are found at various places in Madagascar. It is plentiful in the Kovdor Massif, Russia, where it also forms huge crystals. At Slyudyanka, Lake Baikal district, Russia, it forms large dark red-brown (but surprisingly transparent) crystals with white calcite and granular diopside. Known polytypes: **1M**, **1Md**, 3T, $2M_1$.

Figure 77. Large and very dark phlogopite crystal in a matrix of calcite and green diopside, from Slyudyanka, Irkutsk, Russia. Although the crystal is quite dark, the small backlit area (lower right) shows that it is nevertheless fairly transparent. RJL1311

Figure 78. Crude brown phlogopite crystal about 15 mm tall, in calcite/apatite matrix, from Tory Hill, Ontario, Canada. RJL2105

Figure 79. Phlogopite single crystal showing bronze, submetallic luster on the cleavage surface, from Vohitrosi, Madagascar. RJL709

Figure 80. Transparent, pale greenish phlogopite with a bluish-gray corundum crystal from Mogok, Myanmar demonstrating the reaction dolomite + muscovite = phlogopite + corundum. RJL2620

Polylithionite forms scaly aggregates, cryptocrystalline masses, and sometimes, tabular crystals up to about 5 cm across. It is usually white, cream, or pinkish but may be pale greenish or bluish. Some specimens fluoresce lemon yellow under SW UV. It occurs in syenite pegmatites of the Lovozero massif, Kola Peninsula, Russia, and in the Ilimaussaq alkaline complex, Greenland. It is fairly common at Mont Saint-Hilaire, Quebec, Canada, where UV can be used to distinguish it from similar-looking but non-fluorescent tainiolite (Mandarino and Anderson 1989). It has also been reported from the Dara-I-Pioz massif, Alai Range, Tadjikistan; at Vora, Sandefjord, Oslo region, Norway; and Point of Rocks, Colfax County, New Mexico. Known polytypes: 1M, $2M_1$, 3T.

Figure 81. Paper-thin polylithionite crystals forming a rosette 4 cm across, with an aegirine crystal in the center, from Mont Saint-Hilaire, Quebec, Canada. RJL3194

Figure 82. Silvery foliated masses of polylithionite from Mt. Ploskaja, Kola Peninsula, Russia. Sample is about 15 mm wide. RJL3191

Preiswerkite, the Na analogue of eastonite, forms colorless to pale green platy grains up to about 1 mm, as nodular aggregates with Al-pargasite and zoisite in a rodingite dike within the Geisspfad ultramafic complex, Binntal, Valais, Switzerland (Keusen and Peters 1980). Preiswerkite has also been reported from tourmaline-biotite-scapolite rock at Blengsvaten, Norway. Known polytypes: **2M$_1$**, 1M, 1Md.

Figure 83. Typical white to pale greenish nodule of preiswerkite about 15 mm wide, in matrix from the type locale, Geisspfad, Switzerland. RJL3205

Roscoelite forms fine-grained masses and spherical aggregates of minute scales, typically in certain vanadium-rich sandstones. Roscoelite was first described from the Stuckslacker (Sam Sims) mine, on Granite Creek, near Coloma, El Dorado County, California; it also occurs with native gold near Sutter's Mill, California. It is an important component of the sandstone cement in the "black ore zone" at the Mounana mine, Gabon (Cesbron and Bariand 1975) and is also fairly widespread in sedimentary deposits on the Colorado Plateau. It forms radiating micaceous aggregates and coatings on electrum in epithermal veins at the Engineer mine, Tagish Lake, British Columbia, Canada (Mauthner, Groat, and Raudsepp 1996). Roscoelite has also been reported from the Yamato mine, Japan; the Hemlo gold deposit, Ontario, Canada; and the Kalgoorlie goldfield, Western Australia, Australia. Known polytypes: **1M**, $2M_1$.

Figure 84. Flakes of roscoelite in matrix from the type locale, Stuckslacker mine, California. RJL3172

Figure 85. Olive green foliated rosettes of roscoelite with stringers of a gold-silver alloy (*electrum*) from Engineer mine, Tagish Lake, British Columbia, Canada. Field of view is about 5 mm wide. RJL3187

Shirokshinite, the Na-dominant analogue of tainiolite, was recently described from a pegmatite at the underground Kirovskii apatite mine, Mt. Kukisvumchorr, Khibiny massif, Kola Peninsula, Russia (Pekov et al. 2003). It forms hexagonal prismatic, typically skeletal, crystals and sheaflike structures to several mm. The type mineral is colorless to pale gray or greenish. It occurs on microcline and kupletskite crystals in cavities within the hyperalkaline pegmatite. Later work in a nearby pegmatite uncovered a slightly different habit, in which white to pale tan scaly aggregates are partially replacing biotite (Pekov and Podlesnyi 2004) Known polytype: 1M.

Shirozulite, the Mn analogue of phlogopite, forms minute (0.5 mm) dark reddish brown grains in a regionally and thermally metamorphosed strata-bound manganese ore deposit at the Taguchi mine, Aichi Pref., Honsu, Japan (Ishida, Hawthorne, and Hirotawari 2004). The mineral grains occur in tephroite-rhodochrosite ores in contact with a feldspar vein. It has also been reported from the Bikkulovskoe deposit, South Urals, Russia. Known polytype: 1M.

Siderophyllite is the accepted name for the composition range that includes many of the "zinnwaldites," particularly those found at Pikes Peak, Colorado in association with microcline. Brownish red crystals to 5 cm across are found in pegmatite areas of the Baveno granite, Novara, Italy. Some other occurrences are in the Brooks Range, Alaska; Mont Saint-Hilaire, Quebec, Canada; Altenberg, Saxony, Germany; Reschen Pass, Trentino-Alto Adige, Italy; Badzhala, Ural Mountains, Russia; and Newcastle, County Down, Ireland. Known polytype: 1M.

Figure 86. Macro photo of a mass of brown biotite in the process of partial replacement by tan to silvery shirokshinite, from the type locale at Kirovskii apatite mine, Mt. Kukisvumchorr, Kola Peninsula, Russia. RJL3210

Figure 87. Photomicrograph of the specimen in the previous figure, showing the intimate intergrowth of shirokshinite with the original black biotite. RJL3210

Figure 88. Photomicrograph showing tiny pods of dark reddish shirozulite in black massive matrix, from the Bikkulovskoe deposit, Southern Urals, Russia. RJL3173

Tainiolite (taeniolite) forms colorless to brownish tabular pseudohexagonal crystals to about 5 cm across and 5 mm thick, as well as scaly or crypocrystalline masses in nepheline syenite pegmatites. It was first described in 1901 from the Narssarssuk pegmatite, Greenland. Other localities include: Magnet Cove, Arkansas; Coyote Peak, Humboldt County, California; Mont Saint-Hilaire, Quebec, Canada; in Russia at several localities in the Lovozero massif, Kola Peninsula and at the Burpala alkaline massif in Transbaikalia. Known polytypes: **1M**, **3T**, $2M_1$.

Tetra-ferri-annite forms minute flakes and grains (to about 40 μm) in a very low-grade metamorphosed banded iron formation of the Dales Gorge member, Hamersley Range, Western Australia, associated with hematite, magnetite, quartz, ankerite, stilpnomelane, and riebeckite (Miyano and Miyano 1982). This species is called ferri-annite by some authors. Known polytype: 1M.

Figure 89. Silvery rosettes of tainiolite to about 1 cm across, most with an unidentified black earthy coating, from Mt. Nepkhe-Nelm, Lovozero Massif, Kola Peninsula, Russia. RJL3188

Tetra-ferriphlogopite is reported from the Phlogopite mine in the Kovdor Massif, Kola Peninsula, Russia, where it forms spectacular dark brown crystals up to 10 cm across in massive white calcite (Ivanyuk and Yakovenchuk 1997). It has also been found in an alkaline carbonatite near Tapira, MG, Brazil. Known polytype: 1M.

Tobelite is unusual in that it contains ammonium as the dominant interlayer cation. It is found as a clay-like material with quartz at the Ohgidani pottery stone deposit, Tobe, Ehime Pref., Shikoku, Japan, where it is a hydrothermal alteration product of a biotite andesite dike. It also occurs in an altered rhyolite tuff at the Horo pyrophyllite deposit, Toyosaka, Hiroshima Pref., Honshu, Japan (Higashi 1983). Known polytypes: **1M**, **$2M_1$**.

Trilithionite represents one of the bounding compositions of the lepidolite series. The actual occurrence of natural material of this composition is considered by some authors to be hypothetical. However, material was recently reported from the Neuland quarry, Dobschutz, Saxony, Germany where it forms minute (< 1 mm) pink scaly masses. Other localities include: a lepidolite pegmatite at Rožna, near Bystrice nad Pernstejnem in the Czech Republic; and Sn-Ta mineralized granites of the Monte Alegre de Goias region, Brazil. Known polytypes: 1M, $2M_1$, 3T.

Wonesite occurs as epitactic intergrowths with phlogopite and talc in metamorphosed Post Pond volcanics in the southwestern corner of the Mt. Cube quadrangle, Vermont (Spear, Hazen, and Rumble 1981). Known polytype: 1Md.

Figure 90. Rough black crystal section of tetraferriphlogopite in white massive calcite, from the Kovdor Massif, Kola Peninsula, Russia. RJL1614

Figure 91. Small off-white chalky mass of tobelite from the type locale, Tobe, Japan. RJL3206

Zinnwaldite species are less common than the other trioctahedral micas and typically occur in granite pegmatites and in cassiterite-bearing pneumatolytic veins. Published analyses of zinnwaldite from the type locale in Cínovec, Czech Republic (formerly Zinnwald, Germany) would place it within the compositional field of polylithionite. Slightly more Fe-rich material from Sadisdorf, Germany would properly be called siderophyllite. For the collector, zinnwaldites are occasionally attractive in their own right and often have interesting associates such as cassiterite, wolframite, topaz, scheelite, tourmaline, and fluorite. A classic association from El Paso County, Colorado, is well-crystallized zinnwaldite with blue microcline (amazonite). The Morefield mine, Amelia County, Virginia, has produced many excellent large crystals of zinnwaldite. A number of crystals from the Morefield mine were analyzed and based on the average composition the calculated formula placed these materials nearly midway between the end members, but slightly closer to polylithionite than to siderophyllite (L. E. Kearns, personal communication). On the other hand, individual samples varied significantly around the average composition, so a good case can be made that using the term "zinnwaldite" is perhaps more accurate and descriptive than assigning all Morefield specimens to polylithionite. Known polytypes: **1M**, **1Md** , $2M_1$, 3T.

Figure 92. A small (8 mm) tabular zinnwaldite crystal on matrix, from Rockport, Massachusetts. RJL1343

Micas for the Specialty Collection or Display

Single-species

Some collectors specialize in a single mineral species, usually preferring things like calcite or quartz that show a wide variety of habits, occur worldwide, and are prone to forming attractive specimens. While it would certainly be possible to assemble an interesting collection around a single mica species, a collection devoted to the mica group as a whole can be large and very rewarding. Given the large number of recognized species and the diversity of locales and habits, a collection of suitable depth could easily have hundreds of specimens and many superb display pieces.

Rock types

A collection of pegmatite minerals, for instance, would be incomplete if it didn't feature a good handful of micas and their associates. Micas are equally important constituents in metamorphic deposits, where they sometimes form very large, gemmy crystals, such as the phlogopite crystals in white marble found in several localities, and the bright green crystals of clintonite in blue calcite from the Crestmore quarry, California.

Micas and their associated species

An interesting display case could be assembled on this theme by just about anyone with a large general mineral collection. As an exercise, it might be interesting to go through your collection and examine each piece carefully to see how many specimens of other minerals also have *some* mica present. Matrix specimens of many gem species include mica as an attractive associated species. Excellent specimens of beryl var. aquamarine occur on coarse muscovite in Laghman Province, Afghanistan, and similar specimens of beryl var. heliodor have been found in the Pamirs, Tadjikistan. Pale blue to colorless beryl with a distinctive tabular habit on muscovite from Sichuan Province, China is relatively plentiful at present. Garnet crystals, particularly almandine, frequently occur in mica schist; the hardness of the garnet and softness of the matrix allow "textbook" dodecahedral crystals to be exposed with minimal specimen preparation. Members of the tourmaline group often occur with mica; well-known examples of this association include the lithium-rich pegmatites of California and Brazil. Deep brown, transparent dravite occurs in muscovite at Gujarkot, West Nepal. Other minerals that can form well-defined crystals in mica schist at various localities include staurolite, kyanite, and actinolite.

Figure 93. Blue bladed kyanite crystals in silvery-green muscovite (var. *fuchsite*) schist, from the Kola Peninsula, Russia. RJL1370

Figure 94. Large staurolite crystals, including a well-formed "sixling" twin 6 cm across, in mica schist, from Keivy, Russia. RJL2460

Figure 95. Prismatic crystal of beryl var. *aquamarine* about 3 X 8 cm with muscovite, a typical pegmatite association, from Laghman Province, Afghanistan. RJL1199

Figure 96. Transparent golden crystals of beryl var. *heliodor* with muscovite, from Yasseen, Pakistan. RJL2810

Figure 97. Dark green actinolite crystals in mica schist, from Lake Wenatchee, Washington. RJL1690

Figure 98. A classic example of almandine in mica schist, from Wrangell Island, Alaska. The garnet crystal is about 3 cm across. RJL72

Figure 99. Brown prismatic dravite crystals (largest is 4 cm tall) in muscovite from Gujarkot, Nepal. RJL2580

Figure 100. Gemmy chartreuse brazilianite crystal about 2 cm tall perched on twinned muscovite crystals, from the Duquinha mine, Minas Gerais, Brazil. RJL2845

Figure 101. Clusters of dark green fluorapatite crystals on red-brown muscovite, from Jose Pinto Mine, Minas Gerais, Brazil. RJL2409

Figure 102. An interesting association piece: green octahedral fluorite crystals about 1 cm wide, on sharp silvery muscovite crystals, from Shengus, Pakistan. RJL3156

Figure 103. Dark crystal of phlogopite about 2 cm across and 1 cm thick with pectolite in limestone, from Mont Saint-Hilaire, Quebec, Canada. Note that at the time of purchase this specimen was labeled "muscovite;" however, muscovite is fairly rare at MSH and is not found in limestone blocks, leading to the conclusion that the sample is almost certainly phlogopite. RJL2550

Figure 104. Pale tan phlogopite crystals with red corundum (ruby) in marble from Myanmar; another nice example of the reaction dolomite + muscovite = phlogopite + corundum that occurred during metamorphism. RJL3099

Pseudomorphs

Colorful pseudomorphs of pink lepidolite replacing dark green elbaite occur at Itinga, Minas Gerais, Brazil. Muscovite after nepheline is found east of the Morrison quarry, Dungannon Township, Ontario, Canada. Muscovite after almandine is found at the Bumpus Quarry, Maine. Pink muscovite after elbaite has been found at several localities in Maine, including Hatch Farm Quarry, Auburn, and Mount Rubellite Quarry, Hebron. Muscovite after topaz is found at Lord Hill Quarry, Stoneham, Maine. At the Harding mine, Taos County, New Mexico, muscovite replacements of microcline, spodumene, and topaz have been reported. Pseudomorphs of muscovite partially or completely replacing cordierite are found at Baveno, Italy. In rocks that have been formed by polymetamorphic events, margarite may form pseudomorphs after andalusite, kyanite, sillimanite, corundum, chloritoid, staurolite, or muscovite (Guidotti and Cheney 1976). Bityite after beryl has been reported from Tittling, near Passau, Bavaria, Germany. Spectacular pseudomorphs of muscovite after feldspar and cordierite have been found at the Mokrusha mine, Murzinka pegmatite district, Russia (Popova, Popov, and Kanonerov 2002). The "pseudoleucite" crystals found at Kirshehir, Turkey, present a complicated case: They are occasionally sold as "illite after leucite" or "kaolinite after leucite." XRD studies done at the Czech Geological Survey in May 1999 identified the material as illite, whereas other specimens were found to contain sericitic muscovite and minor amounts of smectite, according to XRD studies reported by Lunel and Akiman (undated paper). This example serves to emphasize the difficulties inherent in characterizing mixtures of extremely fine-grained claylike materials. In addition to pseudomorphs involving replacement of other minerals by mica, cases are also known in which another mineral has developed as a replacement of mica; hematite after phlogopite, reported from Obsidian Cliffs, Lane County, Oregon, is one example.

Figure 105. A spectacular pseudomorph in which pink lepidolite has partially replaced dark green elbaite, from Brazil. Specimen is about 13 cm tall. RJL2387

Figure 106. Detail of the illite casts after halite shown in Figure 52, illustrating how extremely fine illite particles have faithfully preserved the hopper morphology of the original halite crystals. RJL3201

Figure 107. A classic pseudomorph consisting of fine-grained muscovite ("sericite") replacing leucite from Kershehir, Turkey. Specimen is about 6 cm across. RJL3200

Figure 108. A rough pseudomorph of muscovite replacing tourmaline from the McGuiness mine, North Groton, New Hampshire. Specimen is about 8 cm tall.

Fluorescent minerals

Phlogopite crystals from many localities fluoresce dull to bright yellow; the effect is best seen in colorless or transparent specimens (Robbins 1994). Good examples occur at Franklin, New Jersey; Newcomb, New York; Bancroft, Ontario, Canada; Myanmar; and Afghanistan. Lepidolite from a pegmatite near Morristown, Arizona is weakly fluorescent yellow. Polylithionite from Mont Saint-Hilaire, Quebec, Canada; Langesundfjord, Norway; Tadjikistan; and Julianehab, Greenland shows lemon yellow response. Margarite from Franklin, New Jersey fluoresces weak bluish-white and is often associated with other colorful fluorescent species. Bityite may display a dull yellow fluorescence.

Figure 109. A row of tabular, transparent brown phlogopite crystals (each about 15 mm wide X 6 mm thick) in white marble from Badakshan, Afghanistan, shown in natural light. RJL2430

Figure 110. The specimen in the previous figure, shown in SW UV light, where the distinctive yellow fluorescence of the phlogopite crystals can be seen clearly. RJL2430

Further Reading

In addition to the references cited, readers interested in a comprehensive scientific review of the mica group might consider the book, *Micas, Reviews in Mineralogy* Vol. 13, S. W. Bailey, editor, available from the Mineralogical Society of America (1984). This volume of review papers, citing thousands of original references and reports, clearly shows the complexity of the mica group and its importance to the science of petrology. The micas are exhaustively treated in Vol. 3A *Sheet silicates: Micas*, of the Second Edition of *Rock-Forming Minerals* (Fleet 2003).

Ankinovich, S. G., E. A. Ankinovich, I. V. Rozhdestvenskaya, and V. A. Frank-Kamenetskii 1973. Chernykhite, a new barium-vanadium mica from northwestern Karatau, *American Mineralogist* 58: 966 (abstract of paper originally published 1972 in Russian).

Bailey, S. W. 1984. Classification and structures of the micas, *Reviews in Mineralogy* 13: 1-12.

Cairncross, B. and R. Dixon 1995. *Minerals of South Africa*, Johannesburg: The Geological Society of South Africa, 290 pages.

Cassedanne, J. P. and J. Alves 1990. Apatite and muscovite from the Ze Pinto prospect, Minas Gerais, Brazil, *Mineralogical Record* 21(5): 405-8.

Cassedanne, J. and J. Cassedanne 1981. The Urubu pegmatite and vicinity, *Mineralogical Record* 12(2): 73-77.

Černy, P., R. Chapman, D. K. Teertstra, and M. Novák 2003. Rubidium- and cesium-dominant micas in granitic pegmatites, *American Mineralogist* 88: 1832-35.

Černy, P., J. Stanek, M. Novák, H. Baadsgaard, M. Rieder, L. Ottolini, M. Kavalová, and R. Chapman 1995. Geochemical and structural evolution of micas in the Rožná and Dobrá Voda pegmatites, Czech Republic, *Mineralogy and Petrology* 55: 177-201.

Cesbron, F. and P. Bariand 1975. The uranium-vanadium deposit of Mounana, Gabon, *Mineralogical Record* 6(5): 237-49.

Dasgupta, S., S. Chakraborti, P. Sengupta, P. K. Bhattacharya, H. Bannerjee, and M. Fukuoka 1989. Compositional characteristics of kinoshitalite from the Sausar group, India, *American Mineralogist* 74: 200-2.

Deer, W. A., R. A. Howie, and J. Zussman 1962. *Rock-Forming Minerals,* Vol. 3, *Sheet silicates*, London: Longman's, Green and Co., 270 pages.

Eggleton, R. A. and P. M. Ashley 1989. Norrishite, a new manganese mica, $K(Mn^{3+}{}_2Li)Si_4O_{12}$, from the Hoskins mine, New South Wales, Australia, *American Mineralogist* 74: 1360.

Filut, M. A., A. C. Rule, and S. W. Bailey 1985. Crystal structure of anandite-2*Or*, a barium- and sulfur-bearing trioctahedral mica, *American Mineralogist* 70:1298-1308.

Fleet, M. E. 2003. *Rock-Forming Minerals*, Vol. 3A, Second Edition, *Sheet silicates: Micas*, London: The Geological Society, 758 pages.

Foord, E. E., L. B. Spaulding, R. A. Mason, and R. F. Martin 1989. Mineralogy and paragenesis of the Little Three pegmatites, Ramona District, San Diego County, California, *Mineralogical Record* 20(3): 101-27.

Foord, E. E., P. Černý, L. L. Jackson, D. M. Sherman, and R. K. Eby 1995. Mineralogical and geochemical evolution of micas from miarolytic pegmatites of the anorogenic Pikes Peak batholith, Colorado, *Mineralogy and Petrology* 55:1-26.

Frondel, C. and J. Ito 1966. Hendricksite, a new species of mica, *American Mineralogist* 51: 1107-23.

Gaines, R. V., H. C. W. Skinner, E. E. Foord, B. Mason, and A. Rosenzweig 1997. *Dana's New Mineralogy*, New York: John Wiley & Sons, Inc.

Gianfagna, A., F. Scordari, S. Mazziotti-Tagliani, G. Ventruti, and L. Ottolini 2007. Fluorophlogopite from Biancavilla (Mt. Etna, Sicily, Italy): Crystal structure and crystal chemistry of a new F-dominant analog of phlogopite, *American Mineralogist*, 92: 1601-09.

Gnos, E., T. Armbruster, and I. M. Villa 2003. Norrishite, $K(Mn^{3+}{}_2Li)Si_4O_{10}(O)_2$, an oxymica associated with sugilite from the Wessels Mine, South Africa: Crystal chemistry and ^{40}Ar-^{39}Ar dating, *American Mineralogist* 88:189-94.

Goldschmidt, V. 1916. *Atlas der Krystallformen* [see Facsimile Reprint in Nine Volumes (1986) by the Rochester Mineralogical Symposium].

Graeser, S., C. J. Hetherington, and R. Giere 2003. Ganterite, a new barium-dominant analogue of muscovite from the Berisol Complex, Simplon Region, Switzerland, *Canadian Mineralogist* 41: 1271-80.

Guggenheim, S. and H. E. Frimmel 1999. Ferrokinoshitalite, a new species of brittle mica from the Broken Hill mine, South Africa: structural and mineralogical characterization, *Canadian Mineralogist* 37: 1445-52.

Guidotti, C. V. 1984. Micas in metamorphic rocks, *Reviews in Mineralogy* 13: 357-467.

Guidotti, C. V. and J. T. Cheney 1976. Margarite pseudomorphs after chiastolite in the Rangeley area, Maine, *American Mineralogist* 61: 431-4.

Harada, K., M. Honda, K. Nagashima, and S. Kanisawa 1977. Masutomilite, manganese analogue of zinnwaldite, with special reference to masutomilite-lepidolite-zinnwaldite series, *American Mineralogist* 62: 594 (abstract of paper originally published 1977 in Japanese).

Hawthorne, F. C., D. K. Teertstra, and P. Černý 1999. Crystal-structure refinement of a rubidian cesian phlogopite, *American Mineralogist* 84: 778-81.

Higashi, S. 1983. Tobelite, a new ammonium dioctahedral mica, *American Mineralogist* 68:850 (abstract of paper originally published 1982 in Japanese).

Ishida, K., F. C. Hawthorne, and F. Hirotawari 2004. Shirozulite, $KMn_3(SI_3Al)O_{10}(OH,F)_2$, a new manganese dominant trioctahedral mica: Description and crystal structure, *American Mineralogist* 89: 232-238.

Ivanyuk, G. Yu. and V. N. Yakovenchuk 1997. *Minerals of the Kovdor Massif*, Apatity: RAS Kola Science Center Publishing.

Jahns, R. H. and R. C. Ewing 1977. The Harding mine, Taos County, New Mexico, *Mineralogical Record* 8(2): 115-26.

Keusen, H. R. and T. Peters 1980. Preiswerkite, an Al-rich trioctahedral sodium mica from the Geisspfad ultramafic complex (Penninic Alps), *American Mineralogist* 65: 1134-37.

Kogarko, L. N., Y. A. Uvarova, E. Sokolova, F. C. Hawthorne, L. Ottolini, and J. D. Grice 2005. Oxykinoshitalite, a new species of mica from Fernando de Noronha Island, Pernambuco, Brazil; occurrence and crystal structure, *Canadian Mineralogist* 43: 1501-10.

Lauf, R. J. 2006. Collector's guide to the mica group, *Rocks & Minerals* 81(3): 214-24.

Li, G., D. R. Peacor, D. S. Coobs, and Y. Kawachi 1997. Solid solution in the celadonite family: The new minerals ferroceladonite, $K_2FeFeSi_8O_{20}(OH)_4$, and ferroaluminoceladonite, $K_2FeAl_2Si_8O_{20}(OH)_4$, *American Mineralogist* 82: 503-11.

Liang, J.-J., F. C. Hawthorne, M. Novák, P. Černý, and L. Ottolini 1995. Crystal-structure refinement of boromuscovite polytypes using a coupled Rietveld-static-structure energy-minimization method, *Canadian Mineralogist* 33: 859-65.

Ma, C. and G. R. Rossman 2006. Ganterite, the barium mica $Ba_{0.5}K_{0.5}Al_2(Al_{1.5}Si_{2.5})O_{10}(OH)_2$, from Oreana, Nevada, *American Mineralogist* 91: 702-5.

Mandarino, J. A. 1999. *Fleischer's glossary of mineral species 1999*, Tucson: The Mineralogical Record, Inc.

Mandarino, J. A. and V. Anderson 1989. *Monteregian treasures: The minerals of Mont Saint-Hilaire, Quebec*, Cambridge: Cambridge University Press.

Mauthner, M. H. F., L. A. Groat, and M. Raudsepp 1996. The Engineer mine, Tagish Lake, British Columbia, *Mineralogical Record* 27(4): 263-73.

Miyano, T., and S. Miyano 1982. Ferri-annite from the Dales Gorge member iron-formation, Wittenoom area, Western Australia, *American Mineralogist* 67: 1179-94.

Novák, M. and P. Černý 1998. Abundances and fractionation trends of Rb and Cs in micas from lepidolite- and elbaite-subtype pegmatites in the Moldanubicum, Czech Republic, *Acta Universitatis Carolinae – Geologica* 42(1): 86-90.

Novák, M., P. Černý, M. Cooper, F. C. Hawthorne, L. Ottolini, Z. Xu, and J.-J. Liang 1999. Boron-bearing $2M_1$ polylithionite and $2M_1$ + $1M$ boromuscovite from an elbaite pegmatite at Rečice, western Moravia, Czech Republic, *European Journal of Mineralogy* 11:669-78.

Ni, Y. and J. M. Hughes 1996. The crystal structure of nanpingite-$2M_2$, the Cs end-member of muscovite, *American Mineralogist* 81: 105-10.

Odom, I. E. 1981. Glauconite and celadonite minerals, *Reviews in Mineralogy* 13: 545-72.

Pekov, I. V., N. V. Chukanov, E. V. Rumiantseva, Yu. K. Schneider, and N. V. Ledeneva 2001. Chromceladonite – a new mineral of the mica group, *Mineralogical Record* 32(3): 212 (abstract of paper originally published 2000 in Russian).

Pekov, I. V., N. V. Chukanov, G. Ferraris, G. Ivaldi, D. Yu. Pushcharovsky, and A. E. Zadov 2003. Shirokshinite, $K(NaMg_2)Si_4O_{10}F_2$, a new mica with octahedral Na from Khibiny massif, Kola Peninsula: descriptive data and structural disorder. *European Journal of Mineralogy* 15: 447-454.

Pekov, I. V and A. S. Podlesnyi 2004. Kukisvumchorr deposit: Mineralogy of alkaline pegmatites and hydrothermalites, *Mineralogical Almanac* 7, 140 + xxiii pp.

Popova, V. I., V. A. Popov, and A. A. Kanonerov 2002. Murzinka: Alabaska pegmatite field, *Mineralogical Almanac* 5, 128 pp.

Reznitsky, L. Z., E. V. Sklyarnov, Z. F. Ushchapovskaya, N. V. Nartova, V. G. Evsyunin, A. A. Kashaev, and L. F. Suvorova 1998. Chromphyllite – a new dioctahedral mica, *American Mineralogist* 83: 652-3 (abstract of paper originally published 1997 in Russian).

Rieder, M., G. Cavazzini, Y. S. D'yakonov, V. A. Frank-Kamenetskii, G. Gottardi, S. Guggenheim, P. V. Koval, G. Muller, A. M. R. Neiva, E. W. Radoslovich, J.-L. Robert, F. P. Sassi, H. Takeda, Z. Weiss, and D. R. Wones 1998. Nomenclature of the micas, *Canadian Mineralogist* 36: 905-12.

Robbins, M. 1994. *Fluorescence: Gems and minerals under ultraviolet light*, Phoenix: Geoscience Press, 374 pages.

Shen, G., Q. Lu, and J. Xu 2001. Fluorannite: a new mineral of the mica group from the western suburb of Suzhou City, *American Mineralogist* 86:1534 (abstract of paper originally published 2000 in Chinese).

Sinkankas, J. 1964. *Mineralogy for amateurs*, New York: Van Nostrand.

Spear, F. S., R. M. Hazen, and D. Rumble, III 1981. Wonesite: a new rock-forming silicate from the Post Pond Volcanics, Vermont, *American Mineralogist* 66: 100-5.

Speer, J. A. 1984. Micas in igneous rocks, *Reviews in Mineralogy* 13: 299-356.

Yang, Y, Y. Ni, L. Wang, W. Wang, Y. Zhang, and C. Chen 1990. Nanpingite – a new cesium mineral, *American Mineralogist* 75: 708-9 (abstract of paper originally published 1988 in Chinese).

Yoshii, M., K. Maeda, T. Kato, T. Watanabe, S. Yui, A. Kato, and K. Nagashima 1975. Kinoshitalite, a new mineral from the Noda-Tamagawa mine, Iwate Prefecture, *American Mineralogist* 60: 486-7 (abstract of paper originally published 1973 in Japanese).